Magnetomicrofluidic Circuits for Single-Bioparticle Transport

Roozbeh Abedini-Nassab

Magnetomicrofluidic Circuits for Single-Bioparticle Transport

 Springer

Roozbeh Abedini-Nassab
Faculty of Mechanical Engineering
Tarbiat Modares University
Tehran, Iran

ISBN 978-981-99-1701-3   ISBN 978-981-99-1702-0 (eBook)
https://doi.org/10.1007/978-981-99-1702-0

This Springer imprint is published by the registered company Springer Nature Singapore Pte Ltd.
The registered company address is: 152 Beach Road, #21-01/04 Gateway East, Singapore 189721,
Singapore

Preface

This text focuses on the physics and techniques of bioparticle transport in magnetic-based microfluidic systems, called magnetomicrofluidic circuits. It evolved from the recent work me and my colleagues did on these circuits at Duke University and then later at Tarbiat Modares University. It also was complemented with the works of other groups, and became the main material of a graduate course I have taught in recent years, to students from Biomedical Engineering, Electrical Engineering, and Mechanical Engineering. This text brings together materials from several disciplines, including magnetics, fluid mechanics, circuit theory, and surface chemistry, with the ultimate goal of discussing magnetomicrofluidic circuits as a novel bioparticle manipulating method in the lab-on-a-chip systems.

Since this is the first book on this topic, it serves mainly as a reference for researchers in the field, while it is definitely suitable to be used by graduate course instructors. This book starts with a brief review of the available single-particle manipulation methods, including the ones based on microengraving, electrical forces, magnetic forces, acoustic forces, and optical tweezers. I discuss microfabrication techniques used for fabricating the microfluidic chips. The required background in microfluidics theory for studying fluid flows in microchannels is provided. Then, the magnetic theory and the magnetic materials, at the required level for our purpose in this book, are discussed. Then, we are ready to learn the concept of circuit theory and magnetophoretic circuits. I explain various magnetophoretic circuit elements and their different versions. Next, the integrated magnetomicrofluidic circuits are discussed. I explain how the developed circuits are used in order to perform biological tests. Some sample applications of the magnetomicrofluidic chips are provided.

Tehran, Iran   Roozbeh Abedini-Nassab
August 2022

Contents

Symbols

Symbol	Meaning/Definition
ρ	Density
U	Energy
t	Time
P	Pressure
η	Viscosity
R	Radius
A	Cross-sectional area
R_h	Hydraulic resistance
r_h	Hydraulic radius
Per	Perimeter
Q	Volumetric flow rate
L	Length
W	Width
H	Height
f_D	Darcy friction factor
Re	Reynolds number
u	Fluid velocity
μ	Magnetic permeability
Λ	Aspect ratio
D	Electric displacement
ϕ	Electric potential
ε	Electric permittivity
E	Electric field
M	Magnetization
f_{CM}	Clausius–Mossotti (CM) factor
σ	Electric conductivity
Γ	Surface tension
T	Temperature
B	Magnetic flux density

χ	Magnetic susceptibility
H	Magnetic field intensity
ψ	Surface nanoparticle coverage area
μ_0	Magnetic permeability of vacuum
ε_0	Electric permittivity of vacuum
P	Electric dipole moment
C	Capacitance
L	Inductance
δ	Charge density
I	Current
λ	Line charge density
τ	Time constant
ι	Numerical matching coefficient
K	Wave number
ω	Angular frequency
f	Frequency
λ_w	Wavelength
κ	Mobility
α	Cone angle
ν	Shear stress
β	Ratio of particle radius to the gap size

List of Tables

Chapter 1
Introduction

A key goal in the field of lab-on-a-chip is to develop particle-manipulating platforms for controlled transportation, separation, sorting, filtration, trapping, and detection of tiny colloidal particles, droplets, and cells [1–6]. These systems offer multiple fundamental applications in the fields of single-cell biology, cancer biology, immunology, transplantation, and medicine. It is now recognized that the cell heterogeneity hidden in the biological samples is a fundamental phenomenon to be analyzed; however, it cannot be studied by the traditional average-based biological analysis at the bulk level [7]. For example, the field of single-cell analysis (SCA) has recently emerged with the help of high-impact lab-on-a-chip systems, which holds great promise for capturing and analyzing rare but crucially important cellular events in highly heterogeneous cell populations in tissues. SCA tools have substantial applications in studying phenotypic and genotypic cell heterogeneity, which is difficult, if possible, to be analyzed with the traditional bulk cell culture and measurement systems. Hence, SCA tools have attracted attention in various fields ranging from single-cell drug screening to the detection and analysis of rare latently infected cells [8–10], transplanted grafts [11], or cancer cells [12–14].

Cellular heterogeneity within an isogenic cell population can be due to different mechanisms, such as differences in gene regulatory pathways, mutations, environmental signals, signal history, extrinsic noise, or stochastic variations [15, 16]. These variations between the cells are reflected in both genotypic behavior and phenotypic behavior [17–19]. Also, various cell types with different behaviors exist in a tissue. Proteins (e.g., cytokines, chemokines, and growth factors) secreted by these cells carry wealthy information about their origin (*i.e.*, the secreting cell) and their interaction with other cells and the microenvironment. Studying the cell protein secretion profiles at the single-cell resolution allows linking them to the genotypic data obtained from them, a task that can result in a meaningful understanding of cell biology and immunology. This type of study can reveal substantial findings toward realizing the rare but important biological events relevant to the pathogenesis and the human disease outcome.

© The Author(s), under exclusive license to Springer Nature Singapore Pte Ltd. 2023
R. Abedini-Nassab, *Magnetomicrofluidic Circuits for Single-Bioparticle Transport*,
https://doi.org/10.1007/978-981-99-1702-0_1

Human immunodeficiency virus (HIV) infection is one such case, for example, where a few cells determine the fate of the patient. The latently infected cells constitute the HIV viral reservoir, and their eradication is a key goal of the HIV scientific community [20]. Currently, HIV cure research faces two main challenges: (i) the HIV reservoir within a patient is too small to be easily detected with the traditional methods (*i.e.*, about one in a million resting CD4 memory cells is latently infected) [21], and (ii) from this already small reservoir, only a small subpopulation (*i.e.*, < 1%) can be sufficiently reactivated by the cell-stimulating anti-latency agents [22], a necessary task for their immunological recognition and elimination. To overcome these barriers, a suitable device capable to detect these cells and provide a deep mechanistic understanding of the cellular events responsible for the maintenance of the viral latency at the single-cell resolution is needed.

The applications of the SCAs are not limited to infectious diseases. During the last decades, organ transplantation has saved many lives; however, the complicated post-transplantation cellular behavior is not well understood. Analyzing kidney biopsy samples, for example, has revealed important information, but the associated complications, scoring variability, invasiveness, and costs are some main concerns in the transplantation field. Also, studying the graft dynamic behavior post-transplantation is needed; however, this is considered a challenge, because repeated biopsies are not convenient and practically possible [23]. Thus, other sources of organ injury biomarkers, such as urinary cells and peripheral blood mononuclear cells (PBMCs) from kidney transplant recipients, are considered to be analyzed using the SCA methods. Transcriptomic analysis of these cells at the single-cell level is a noninvasive method that can result in biomarker detection and understanding of the underlying phenomena. Researchers at Weill Cornell Medical College have performed single-cell RNA-sequencing on urinary cells obtained from kidney transplant recipients, with both normal and graft rejection signs [24]. They have reported increased macrophages, dendritic cells, T cells, and NK cells in the acute T-cell-mediated rejection samples, while the normal samples displayed dominant kidney tubular epithelial cells. This work and similar studies have shown that SCAs are innovative tools to uncover the complex cellular landscape in the allograft rejection at the single-cell resolution.

Another example of cellular heterogeneity is seen in cancer, where scientists, to understand the important signaling pathways and events in tumor initiation, progression, metastasis, recurrence, and responses to therapeutic agents, need to analyze the variations at the single-cell resolution [25]. Currently, there are two models explaining the heterogeneity seen within the tumor cells. The first one is the cancer stem cell (CSC) model, where CSCs are hypothesized to form a small subpopulation having major tumorigenic potential [26, 27]. Based on this model, because of the epigenetic mutations in rare CSCs, phenotypically diverse cancer cells with no (or less) tumorigenic potential are formed. Then, these cells produce the tumor tissue [28]. In the clonal evolution model (CE), the cancer cells experience genetic and epigenetic mutations over time. These mutated cells form different sub-clones and again divide, mutate, and form the next sub-clones [28, 29]. The formation of these sub-clones during cancer development and the difference between them in the CE model results in cell heterogeneity. All the resulting cells at different levels contribute to tumor maintenance [30].

SCA tools are considered important tools for uncovering the cell behaviors at single-cell level and for finding the right model in a tumor [29, 31]. For example, SCA detects pre-malignant CSCs and the ones resisting treatments [31]. Also, it is capable of detecting the small sub-clones and identifying the clonal architecture. It can discover the mutation orders in different sub-clones and identify the mutated genes within cancer cell populations. These studies provide the opportunity to plot the single-cell resolution version of the tumor phylogenetic tree. Finally, it can detect the cell-cell interactions in the highly heterogeneous cancer cell populations. These fundamental applications are crucial in the diagnostics, study, and treatment of cancer. Because achieving these goals based on traditional bulk measurements is not easy, scientists have focused on designing and developing novel SCA tools with the mentioned capabilities. Advanced understanding of rare cellular behavior and signaling in such diseases based on with the help of SCAs moves the field toward development of novel therapeutic methods with the ultimate curing goals.

The suitable SCA tools for answering the requirements in this field must offer a proper combination of programmability, flexibility, and scalability at the scales adequate to analyze the bio-events. Toward this goal, a novel, programmable, and massively-parallel SCA tool, called the magnetophoretic circuit, based on the principles of computer circuits, is developed. By adding these magnetic circuits into microfluidic channels, the magnetomicrofluidic platform, with the ability to organize numerous single particles in an array in a controlled manner, is developed.

Magnetophoretic circuits use passive and active circuit elements. The passive elements are constructed in magnetic thin films patterned to move cells and particles along programmed tracks when exposed to an external rotating magnetic field. Cell and particle motion along these magnetic tracks is similar to the charge motion in electrical conductors and follows a rule analogous to Ohm's law. Also, asymmetric conductors are engineered to operate similarly to electrical diodes. These elements transport the particles unidirectionally. Magnetophoretic capacitors operate as particle storage sites. Moreover, magnetophoretic active circuit elements are developed with the use of an overlaid microwire to switch single particles between different magnetic tracks. This switching mechanism is similar to the operation of electronic transistors. It is based on introducing a semiconducting gap in the magnetic track to be tuned from insulating state to conducting state by applying a suitable electrical current to an overlaid electrode (i.e., the transistor gate). An extensive study is performed on the operation of transistors to optimize their geometry and minimize the required gate currents.

By integrating these circuit elements, devices that are capable of organizing a precise number of cells into individually addressable array sites, similar to how a random access memory (RAM) stores electronic data, are built. These programmable magnetic circuits allow for the organization of both cells and single-cell pairs into large arrays. If required, it is possible to retrieve single cells for downstream next-generation genomic analysis.

To enhance the efficiency of the device and to achieve higher particle delivery speeds, microfluidic systems are combined with magnetophoretic circuits. The resulting hybrid system is called a magnetomicrofluidic chip which is capable of rapidly assembling particle and cell arrays in a highly controllable manner. Cells can grow inside these chips during multiple days, for long-term phenotypic analysis of important rare cellular events. These types of studies can reveal important insights into the intercellular signaling networks and answer crucial questions in biology and immunology.

In this text, I will provide the required background and the fundamental concepts required for entering the single-cell analysis field, with a focus on magnetophoretic circuits, as a novel example of SCA tools. Other applications of the magnetomicrofluidic chips will also be discussed.

1.1 Particle and Droplet Manipulation Techniques

To study the bioparticles and droplets, we need to move them to specific spots, where further preparation and analysis are performed. Towards this goal, various methods are currently introduced, which in general can be divided into two main categories of (i) flow-based systems and (ii) array-based systems.

1.1.1 Flow-Based Systems

In flow-based systems, the particles or droplets move in a fluid flow to be sorted or analyzed. In sorter systems, typically similar particles are grouped, and then, different groups move to different microchannels. For example, as shown in Fig. 1.1, using magnetic or acoustic forces (depicted by "Separating Signal" in Fig. 1.1), the particles move laterally in the main microchannel, and then downstream the channel they move to the subdivided microchannels. In Fig. 1.1, the mixture of particles (black and red circles) is flown into the main channel from the inlet, and then, each group is separated and flown into the distinct outlets.

1.1.1.1 Acoustic Wave-Based Particle Sorting Systems

Acoustic waves are mechanical vibrations traveling through solids, liquids, or gases [32]. There are two typical acoustic waves, namely, surface acoustic waves (SAW) and bulk acoustic waves (BAW) [33]. SAWs propagate at the medium surface and usually are created by transforming the radiofrequency (RF) signals into mechanical vibrations. Interdigital transducers (IDT) on piezoelectric substrates can produce these waves (see Fig. 1.2). The application of the radiofrequency signals to the IDTs results in mechanical displacement with waveforms which depend on the electrode

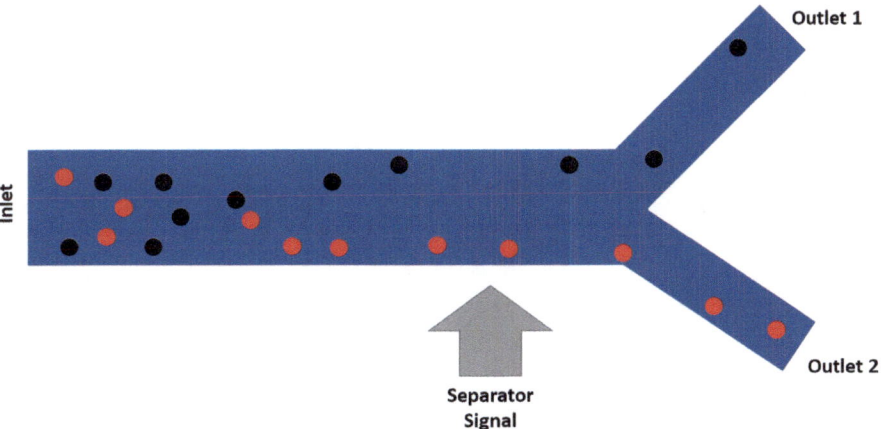

Fig. 1.1 Schematic of a sample flow-based particle separation microfluidic chip is illustrated. Two different sets of particles are shown with the black and red circles

geometries and dimensions, the sound propagation speed in the media, and the input electrical signal power. BAWs propagate in the bulk material and usually are created using a piezoelectric film between two electrodes [34].

The acoustic waves can be used in manipulating particles and cells, as a simple, label-free, and noninvasive technique. In this method, the standing surface acoustic wave (SSAW) creates pressure nodes [35, 36], the positions of which are defined by the microchannel geometry [37] and the particle size. As shown in Fig. 1.2, the particles in a fluid flow move to the produced nodes, and so they can be size-separated [38]. The design introduced in Fig. 1.2 first concentrates the particles in the middle of the channel and then separates them based on their size.

1.1.1.2 Flow Cytometry

In another widely used particle sorting and analyzing method, called flow cytometry, the particles or cells are flown in the device and exposed to laser(s). The light scattered from the particles which carry their information is detected. Two different detectable scattered lights are the forward scatter (FSC), which indicates the relative particle size, and the side scatter (SSC), which indicates the particle structure (e.g., the internal complexity or granularity of the cells or particles). Also, in a specialized version of the flow cytometers, called fluorescence-activated cell sorting (FACS), fluorescent characteristics of each particle or cell are obtained. In this method, before injecting the cells, the samples are prepared for fluorescence measurement by means of transfection and expression of fluorescent proteins, staining with fluorescent dyes or fluorescently conjugated antibodies [39]. Although information of hundreds of proteins and antibodies inside the cells and on their surface, cell counts, and many

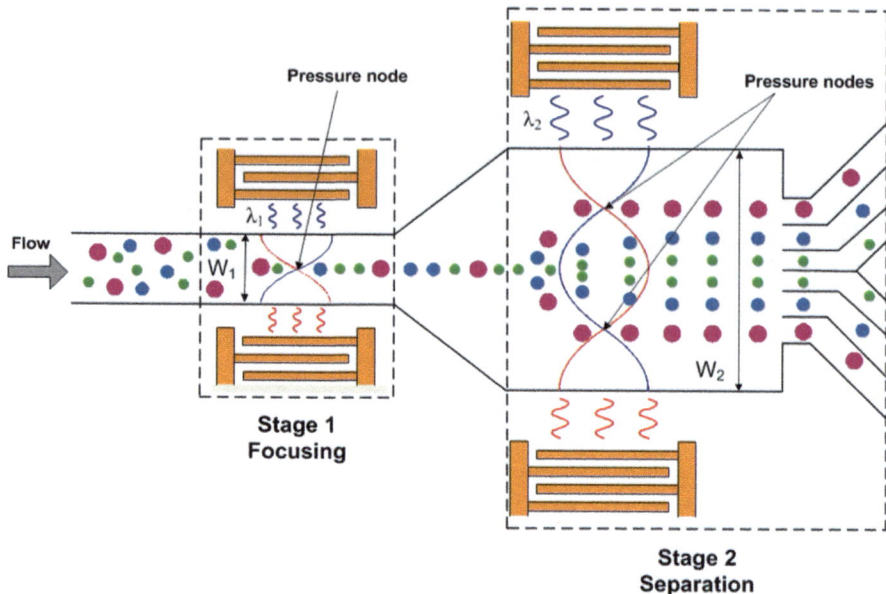

Fig. 1.2 A schematic for a sample particle separator based on standing surface acoustic waves is shown. It is composed of two stages, including the first one for aligning the particles on the center line and the second one for separating them based on their size. The figure is taken from [38] with permission, Copyright 2012 MDPI, under the terms and conditions of the Creative Commons Attribution license (http://creativecommons.org/licenses/by/3.0/)

other chemical and physical data can be extracted, and these systems have led scientists to outstanding achievements [40–42], they are not able to track the dynamic genotypic or phenotypic behavior of single cells over extended periods of time. Moreover, the cells are treated with different chemicals to block their cytokine secretion and/or permeabilize their membrane. Thus, live cell recovery is not possible, and the cell RNA content is degraded, inhibiting their immune transcriptional profiling [43]. Although FACS detects the cytokines associated with the cell membranes [44], the majority of cytokines are secreted from the cells and can thus be ignored by the system. These limitations have motivated researchers to develop other single-cell analysis tools.

1.1.1.3 Droplet-Based Microfluidics

Droplet-based microfluidics is a relatively new device capable of producing a large number of particles in a short time [45–47]. These chips have a wide range of applications, including drug delivery [48], drug screening [49], biomedical imaging [50], diagnostics [51], particle synthesis [52], and single-cell analysis, which is of our interest. Various designs for the droplet-based microfluidic chips are introduced (see

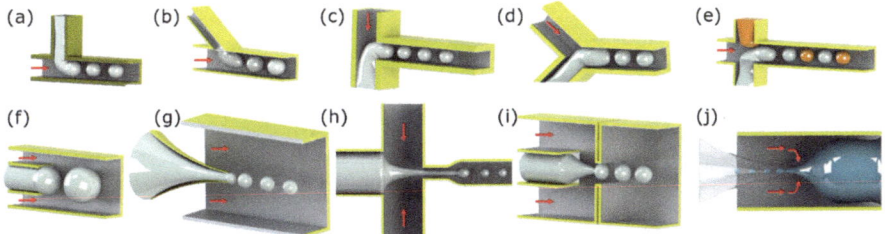

Fig. 1.3 Schematics of droplet-based microfluidic designs are illustrated. (a–e) Cross-flow, (f, g) Co-flow, and (h–j) Flow-focusing. Here, the blue fluid (and the orange fluid in (e)) represent the dispersed phase, and the red arrows depict the direction of the continuous phase flow. The figure is obtained from [52] under the terms and conditions of the Creative Commons Attribution (CC BY) license (https://creativecommons.org/licenses/by/4.0/)

Fig. 1.3), including (i) cross-flow [53–56], (ii) co-flow [57, 58], and (iii) flow-focusing [45, 59–61]. In this method, droplets of the dispersed phase(s) are produced in a continuous phase.

In the cross-flow design (e.g., the T-junction, Y-junction, etc.), the channels containing the two phases make an intersection at which the dispersed phase is cut, based on which the droplets are produced in the continuous phase and transported towards the outlet (see Fig. 1.3a–e). As illustrated in Fig. 1.3e, two different particle sets can also be developed from different dispersed phases, merge afterward, and form more complicated structures. In the co-flow design, the continuous phase symmetrically surrounds the dispersed phase, while they move in coaxial microchannels (see Fig. 1.3f, g). In the flow-focusing configuration, the continuous phase symmetrically shears the dispersed phase in the middle and pushes it through an orifice (see Fig. 1.3h–j). Different forms of flow-focusing design are already introduced. In a simple method, the microchannel containing the dispersed phase meets channel pair(s) on the sides (see Fig. 1.3h). One side channel pair carries the continuous phase, while extra side channel pairs can also bring in other dispersed phases (e.g., for producing multicore or multi-shell droplets). In another version, called 3D flow-focusing, a glass micro-capillary is used to provide the required orifice. Even more complicated versions of this device are introduced, where two glass micro-capillaries are used to form a combination of flow-focusing and co-flow design (see Fig. 1.3j). In these methods, better control of the droplet parameters is achieved [52].

In the droplet-based microfluidics used as a single-cell analysis tool, single cells are encapsulated in the produced nanoliter- or picoliter-sized droplets [62, 63], which are considered tiny microchambers. A schematic of this tool, based on the flow-focusing design, is illustrated in Fig. 1.4. In this method, the required items suspended in the dispersed phase(s) meet at an intersection. The single cells are flown into the junction from channel 3, while other particles (e.g., other cells or the barcode-carrying beads) join from channel 1, and the required buffers contribute from channel 2. At the next junction, the continuous phase (oil) comes in from channels 4 and 5 in Fig. 1.4 and forms droplets out of the dispersed phase containing the mentioned particles

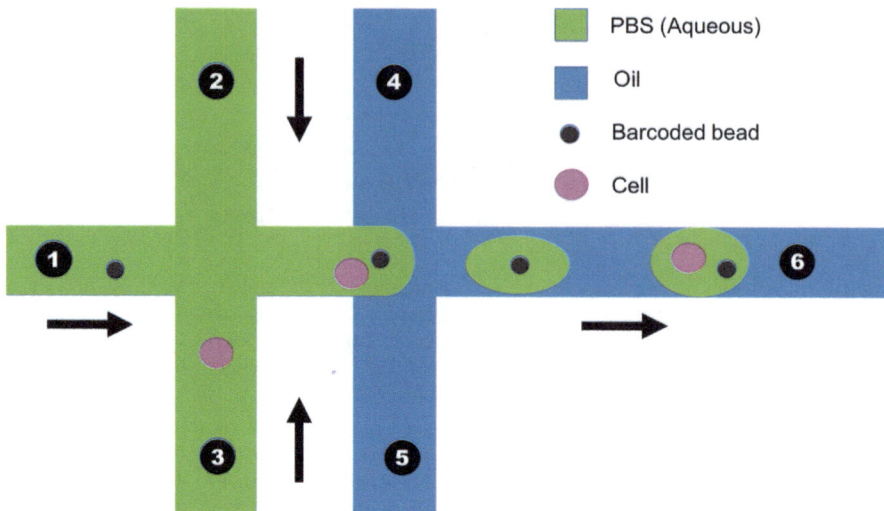

Fig. 1.4 Droplet-based microfluidics schematic for encapsulating cells and beads is shown. Single cells (from channel 3), the other particles (from channel 1), and the required buffers (from channel 2) meet at an intersection and form droplets at the second intersection, where oil plays the role of the continuous phase. The black arrows show the fluid flow direction in channels

and materials. The encapsulated cells in aqueous droplets move through channel 6. In this technique, one needs to keep the cell and bead concentrations low enough to prevent cell or bead doublet formation in the droplets. But low concentrations of cells or beads result in some useless droplets, in which cells or beads are not available. This challenge has been answered with different methods, which is out of the scope of this text.

Encapsulating single cells in droplets has various high-impact applications in biology and medicine. For example, in a method called "Drop-seq," each cell in each droplet is lysed, and its mRNAs are captured by their companion barcode primer-carrying bead in the droplet. Then, the droplets are broken, and after reverse transcribing and amplifying the mRNAs from all single cells, next-generation sequencing tools are used to sequence them at once. Finally, using the barcodes, the cell of origin and gene of origin for any mRNA are identified [64]. In Drop is a similar technology in which hydrogels carrying photocleavable barcode primers are used to capture the mRNAs from the cells in the droplets [65]. The advantage of this method is the ability to increase the hydrogel concentration to extremely limit the number of droplets lacking hydrogels.

In droplet-based microfluidic chips, the reaction volumes (*i.e.*, nanoliters or pico-liters) are typically much lower than the ones in the traditional methods, which leads to faster detection of the secreted proteins from the single cells in the droplets compared to the commonly used screening plates. Also, these systems can produce a large number of droplets in a short time, which means they are high throughput in nature. Hence, in addition to transcription profiling of single cells, droplet-based microfluidic systems are suitable tools for detecting single-cell cytokine secretions [66–68], single-molecule polymerase chain reaction (PCR) [69, 70], single-cell viral infectivity assays [71], rapid isolation of individual cells [72], and other single-cell biological applications. For example, droplet-based microfluidic chips are equipped with a droplet-arraying technique to enable dynamic studies of single-cell protein secretion profiles [67, 73]. In this method, single cells are encapsulated in droplets, which then play the role of tiny chambers for the cells to grow. Although the particle production step in this method is similar to the flow-based techniques, they fall in the array-based system category because of forming arrays.

1.1.2 Array-Based Systems

The flow-based systems, some of which are reviewed in the previous section, are usually high-throughput analysis tools capable of analyzing numerous particles in a short time. But they mostly do not offer to study the dynamic behavior of single cells over time. Array-based systems are considered a solution to overcome this challenge. Multiple methods for assembling single cells into microchambers on the chip to be studied are proposed. Microengraving [74–76] is a simple method for putting numerous single cells and/or cell pairs into microwells. Then, biostudies such as real-time fluorescent analysis in response to external stimuli or probing cell-cell interactions can be performed. It is also possible to retrieve the cells from the microwells by pipetting or by automated micromanipulators [77] to perform follow-up off-chip studies. However, during the sedimentation time, no control over individual particles and cells is offered. Thus, this technique is not suitable for applications such as assembling a certain number of particles in each well or forming pairs of specific particles.

1.1.2.1 Hydrodynamic Trapping

Hydrodynamic trapping of single particles in an array in a microfluidic chip is another widely used method for cell studies and lab-on-a-chip applications [78–82]. This idea is well explained based on the concept of hydraulic resistances, which requires a little bit of background in the field of microfluidics. We start with the description of the pressure-driven flows through a channel of circular cross-section, called Hagen–Poiseuille flows. The governing equations in this type of system are the Navier–Stokes equations, which, assume the flow to be only in the z-direction, and can be written as Eq. (1.1) [83].

$$\rho \frac{\partial \vec{u}}{\partial t} + \rho \vec{u} \cdot \nabla \vec{u} = -\nabla p + \eta \nabla^2 \vec{u}, \tag{1.1}$$

where $\rho, \boldsymbol{u}, t, p$, and η are the density, flow velocity, time, pressure, and fluid viscosity, respectively. Now, assuming flows to be radially symmetric, for a channel of radius R, the solution of Eq. (1.1) for the velocity in cylindrical coordinate can be written as Eq. (1.2).

$$u_z = -\frac{1}{4\eta}\left(R^2 - r^2\right)\frac{\partial p}{\partial z}. \tag{1.2}$$

The volumetric flow rate is obtained by spatial integration of Eq. (1.2) as follows:

$$Q = -\frac{\pi R^4}{8\eta}\frac{\partial p}{\partial z}. \tag{1.3}$$

For a circular channel with length L, by replacing $-\frac{\partial p}{\partial z}$ with $\frac{\Delta p}{L}$, Eq. (1.3) can be approximated as

$$Q = -\frac{AR^2}{8\eta L}\Delta p, \tag{1.4}$$

where A is the channel cross-sectional area. Now, we can define the hydraulic resistance R_h by Eq. (1.5).

$$R_h = -\frac{8\eta L}{AR^2}. \tag{1.5}$$

And we can write

$$Q = \frac{\Delta p}{R_h}, \tag{1.6}$$

which is basically similar to Ohm's law in electrical circuits, where Q and Δp are replaced with the electrical current (I) and electric potential (V), respectively. Equation (1.5) is defined for the microchannels with circular cross-sections. But it can also be used for channels with other cross-sections with a good approximation. For this purpose, we re-write this equation as Eq. (1.7).

$$R_h \approx -\frac{8\eta L}{Ar_h^2}, \tag{1.7}$$

where r_h is called the hydraulic radius of the microchannel and is defined by Eq. (1.8).

$$r_h = \frac{2A}{Per}, \tag{1.8}$$

where *Per* is the perimeter of the channel cross-section. For example, most of the time, the microchannel has rectangular cross sections, and Eq. (1.8) allows us to use Eq. (1.7) in predicting the hydraulic resistance.

To engineer a trap, two fluidic paths are placed in parallel. The ones with relatively lower and higher initial hydrodynamic resistivities are called the trap and the bypass channels, respectively. The fluid and the particles tend to move through the trap path with lower fluid resistivity. But the trap width is designed to be smaller than the particle diameter and thus it captures the incoming particle. Capturing the first particle leads to higher trap hydrodynamic resistivity relative to the bypass channel, and thus the next particle travels into the bypass channel.

Alternatively, we can consider the ratio of the volumetric flows in the parallel channels. Darcy–Weisbach equation is used to determine the pressure difference, and using the Hagen–Poiseuille flow problem, the pressure difference is obtained [84].

$$\Delta p = \frac{f_D L \rho u^2}{4 r_h},\tag{1.9}$$

where f_D is the Darcy friction factor, L is the channel length, and u is the average fluid velocity, respectively. The Darcy friction factor is related to the channel aspect ratio (which is defined such that it always is in the range of $0 \leq \Lambda \leq 1$) and Reynolds number, which is defined by Eq. (1.10).

$$Re = \frac{\rho u L}{\eta}.\tag{1.10}$$

The product of the Darcy friction factor and Reynolds number is defined by a constant $C(\Lambda)$ that depends on the aspect ratio. Hence, after some replacements, the pressure difference is calculated as Eq. (1.11) [84].

$$\Delta p = \frac{C(\Lambda) \eta L Q Per^2}{32 A^3}.\tag{1.11}$$

Now, by writing Eq. (1.11) for the trap and the bypass channels, considering equal pressure drops, the flow ratio is derived as Eq. (1.12) [84, 85].

$$\frac{Q_t}{Q_b} = \left(\frac{c_b(\iota)}{c_t(\iota)}\right)\left(\frac{L_b}{L_t}\right)\left(\frac{W_b + H_t}{W_t + H_b}\right)^2\left(\frac{W_t H_t}{W_b H_b}\right)^3,\tag{1.12}$$

where W and H are the channel width and height, respectively. Indices t and b stand for the trap and the bypass channels, respectively. The ratio in Eq. (1.12) needs to be greater than 1 to have the trapping system working properly. For the trap channel to capture the particles, its width (*i.e.*, W_t) needs to be smaller than the diameter of the particle. Since the channels have the same heights, to achieve the required ratio of volumetric flows, we need to design a bypass channel longer than the trap

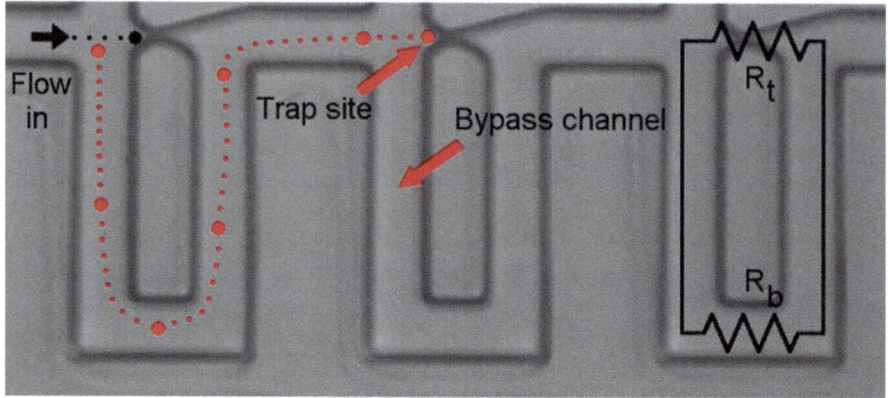

Fig. 1.5 A microscopy image of the single-cell hydrodynamic trapping system is shown. The small black arrow on the left depicts the fluid flow direction toward the microfluidic channel. The little black dots stand for the trajectory of the first entering particle, being caught in the first trap site. The little red dots show the trajectory of the second particle, captured in the second trap site. The position of the big dots is obtained from the experiments; however, the dotted lines are drawn only to help the eyes in tracking the particle trajectories. On the right, the equivalent circuit model for the hydrodynamic channels is drawn. *Rt* and *Rb* are the hydraulic resistances of the trap and bypass channels, respectively. © 2019 IEEE. The Figure is taken with permission from [86]

channel (*i.e.*, $L_b \gg L_t$). Figure 1.5 illustrates a microscopy image of the mentioned trapping system. The black and red dots in this figure depict the trajectory of the first and second entering particles which are caught in the first and second traps. The equivalent circuit model for the trapping system is also drawn.

The trap block shown in Fig. 1.5 is repeated in series to form a chip for entrapping single particles in individual traps. These fluidics traps, for example, can be used to form arrays of single cells to be further studied on the chip.

In addition to the trapping system shown in Fig. 1.5, other particle trapping systems are also introduced. For example, Fig. 1.6 illustrates a schematic of another fluidics trap design in which a cell suspension at high cell concentration is flown into the microchannels [81]. The trapping sites capture the moving cells, and similar to the other method discussed above, after trapping the first particle, other particles are prevented from entering the same site. The hydrodynamic single-particle trapping method is commercialized (Fluidigm® C1). The device offers to analyze 800 single cells using real-time PCR.

A similar technique has also been used to form cell pairs [85, 87]; however, its reported efficiency is as low as ~70%. Also, since no control over individual particles is offered in this technique, another associated challenge is to assemble a cell pair from two particular particles.

Another important task is to retrieve a particle from the assembled array. Towards this goal, researchers have used a laser beam to produce microbubbles at the trap sites. The bubbles push the particle away from the trap, which then moves inside the microchannel to the outlet. Further studies to evaluate the produced heat, forces, and

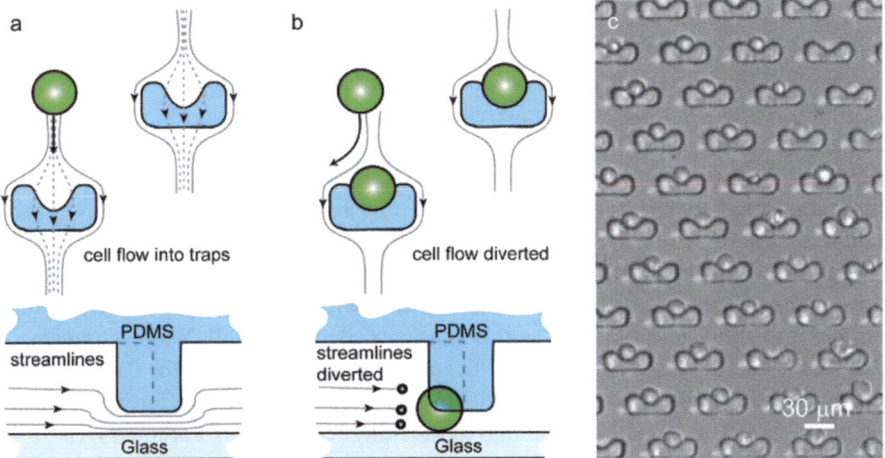

Fig. 1.6 A sample hydrodynamic cell trapping system. (a, b) Schematic of the cell trapping mechanism. (c) A microscope image of the cells trapped in an array. Reprinted with permission from [81]. Copyright 2006 American Chemical Society

their effect on the retrieved cell and its viability, a problem associated with typical laser-based methods, need to be done [88, 89].

The techniques discussed above are based on the fluid flow inside the microchannels and are called passive methods. But particle manipulating approaches based on a force other than hydrodynamic forces are called active methods. The driving force in these systems can be due to the electric field [90–93], optical tweezers [94–96], acoustic forces [97–100], and magnetic fields [2, 101–104].

1.1.2.2 Dielectrophoresis-Based Systems

Dielectrophoresis (DEP) is one of the particle manipulation methods based on electric fields. The techniques based on the DEP forces move electrically polarizable matter in inhomogeneous electric fields. Pohl first introduced the idea in the 1950s [105], but then it has been developed and found several applications in the separation and purification of inorganic [106–109], organic [110–115], and living materials [116–118]. This method has been widely used in microfluidic systems because it is considered a label-free method and the produced forces in micrometer-sized microchannels are large enough to move the particles. Numerical calculation of the DEP forces on various particles has been performed by researchers. For example, the DEP forces acting on a straight, slender body [119] and a chain of spherical particles are proposed [120].

In electrophoresis, the charged particles respond to the applied electric field; however, in DEP, no net charge on the particles is needed. In DEP, Coulomb forces act on the electrically polarizable particles exposed to a non-uniform electric field.

A spherical, uncharged, dielectric particle with electrical permittivity (dielectric constant), ε_p, exposed to an electrical field, gets polarized. Positive and negative charges will form on the opposite sides of the particle. In a uniform electric field, the Coulomb forces on either side are equal and opposite, resulting in a zero net force on the particle. However, in a non-uniform electric field, the sphere feels a larger attractive force on the side with the larger electric field, and thus, it moves in that direction. This behavior (*i.e.*, particle motion towards the regions with high electric fields) is called positive dielectrophoresis (pDEP).

Till now, we have not considered the effect of the surrounding medium (fluid). Now, in the case of a particle suspended in a fluid with electrical permittivity ε_f, both particle and fluid get polarized, and a charge, called Maxwell–Wagner interfacial charge, is generated at the interface of the particle and the fluid. In the case of a non-uniform electric field, this charge results in a net force. Here, the difference between the polarization of the particle and that of the fluid defines the net applied DEP force on the particle. The particle with larger polarization, compared to the fluid, feels forces towards the regions of higher electric fields (*i.e.*, pDEP). But, the particle with less polarization, compared to the fluid, experiences forces towards the regions of lower electric fields, a case called negative dielectrophoresis (nDEP). Please note that in the discussions above, we have assumed the particles and the surrounding fluid to be uniform and perfect dielectrics. In conditions other than that, other effects are involved, which are not discussed here.

To calculate the DEP force, we start with the Laplace equation.

$$\nabla \cdot \vec{D} = -\tilde{\varepsilon}\nabla^2\varphi = 0, \tag{1.13}$$

where $\boldsymbol{D}$ and φ stand for the electric displacement and electric potential, respectively. Here, $\tilde{\varepsilon} = \varepsilon + \sigma/j\omega$ is the complex permittivity, which is frequency dependent. Since the applied electric field can be an AC field, it can be expressed as $\vec{E}_{ext} = \vec{E}_0\hat{z}e^{j\omega t}$ The boundary conditions at $r = a$ and $r = \infty$ can be written as

$$\varepsilon_m\frac{\partial\varphi}{\partial n}\Big|_m - \varepsilon_p\frac{\partial\varphi}{\partial n}\Big|_p = q'' = 0, \tag{1.14}$$

$$\varphi = -E_0 z e^{j\omega t}, \tag{1.15}$$

where indexes f and p stand for the fluid and particle, respectively. The solution is derived in terms of Legendre polynomials. This solution inside and outside the spherical particle is written as Eqs. (1.16) and (1.17).

$$\varphi = -\frac{3\varepsilon_f}{\varepsilon_p + 2\varepsilon_f}E_0 r\cos\theta e^{j\omega t}, \ (r < a) \tag{1.16}$$

$$\varphi = -E_0 r\cos\theta e^{j\omega t} + \frac{\varepsilon_p - \varepsilon_f}{\varepsilon_p + 2\varepsilon_f}E_0\frac{a^3}{r^2}\cos\theta e^{j\omega t}, \ (r > a). \tag{1.17}$$

The first and second terms in Eq. (1.17) correspond to the applied electric field and the point dipole, respectively. The dipole moment of the spherical particle with radius a is derived as Eq. (1.18).

$$P = 4\pi \varepsilon_m E_0 a^3 \frac{\varepsilon_p - \varepsilon_f}{\varepsilon_p + 2\varepsilon_f} \hat{z}. \tag{1.18}$$

In Eq. (1.18), the frequency-dependent Clausius–Mossotti (CM) factor is an important term defined by

$$\tilde{f}_{CM} = \frac{\tilde{\varepsilon}_p(\omega) - \tilde{\varepsilon}_f(\omega)}{\tilde{\varepsilon}_p(\omega) + 2\tilde{\varepsilon}_f(\omega)}. \tag{1.19}$$

This factor is limited in the range of $-0.5 \gg K(\omega) \gg 1$. The time-averaged DEP force then can be calculated as Eq. (1.20).

$$\left\langle \vec{F}_{DEP} \right\rangle = \pi \varepsilon_m a^3 Re\left(\tilde{f}_{CM} \right) \nabla \left(\vec{E}_0 \cdot \vec{E}_0 \right). \tag{1.20}$$

If we consider the spherical particle as a point dipole, its potential energy in this system can be approximated as Eq. (1.21) [121].

$$U_e = -\frac{3}{2} \varepsilon_0 V_p Re[f_{CM}] \vec{E}^2, \tag{1.21}$$

where V_p is the particle volume. Equation (1.21) shows that based on the DEP forces, when the CM factor is positive (negative), the particles move towards the local electric field maxima (minima). Since the CM factor is a function of particle properties and the operating frequency, the DEP forces can be used to separate various particles. The user can define the operation regime simply by adjusting the settings of the used function generator. Also, because the polarity does not play a role in DEP, AC or DC electric fields can be used.

Several electrode geometries have been used for producing the DEP forces. Mathematical modeling of the force produced with interdigitated electrode geometry is proposed [122]. One of the most popular ones is based on the IDT arrays, capable of producing strong electric fields throughout the media [123–125]. The IDT electrode shapes include straight bars, sawtooth [126, 127], square traps [128], quadrupole traps [129, 130], and octupole traps [131, 132]. These systems have been excited using stationary signals [133], traveling waves [134], pulsed signals [135], and multiple frequency signals [136, 137].

The simplest possible electrode geometry, as shown in Fig. 1.7, consists of a rectangular electrode array. If excited with electric potential, as shown in Fig. 1.7c, it produces an energy distribution, with energy minima above the electrodes. This energy simulation shows that this DEP-based device levitates the particles in the nDEP mode, and the particles move toward the substrate in the pDEP mode.

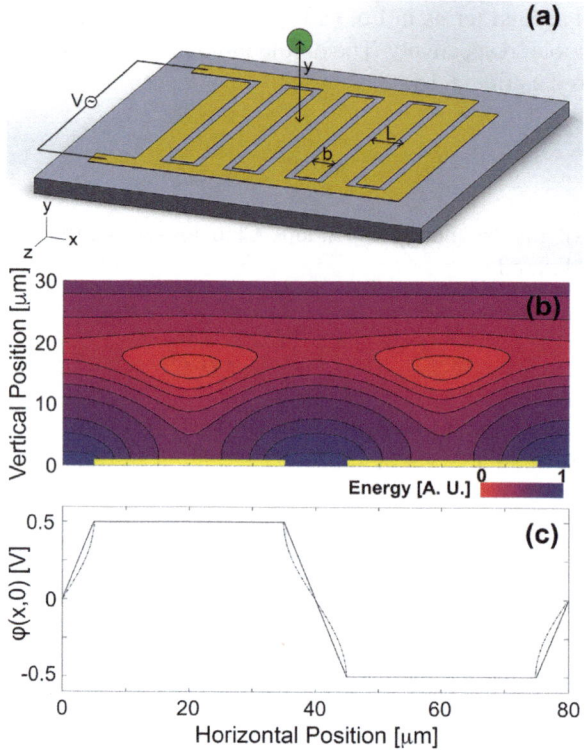

Fig. 1.7 A schematic of a sample nDep chip and potential simulation is shown. (a) The nDep chip layout is illustrated. The nDEP force levitates the particle to a height, y, above the substrate. (b) A sample potential energy landscape over the chip for an AC potential applied to the chip is shown. The red and purple regions stand for energy minima and maxima, respectively. (c) The solid and dashed lines stand for the assumed boundary conditions for the analytical solutions and the simulation results. Reprinted with permission from Springer Nature: Springer Nature, Microfluidics and Nanofluidics, Quantifying the dielectrophoretic force on colloidal particles in microfluidic devices, Abedini-Nassab et al., COPYRIGHT (2022) [121]

To find the electric potential, starting with the Laplace equation ($\nabla^2 \varphi = 0$) and the potential boundary conditions defined on the electrodes as $\pm(V_0/2)\,sin(\omega t)$, and assuming thin electrodes and linear and piece-wise continuous potentials (Fig. 1.7c), one can use the Fourier series [121],

$$\varphi(x, y) = \frac{4V_0}{\pi^2}\left(\frac{L}{L-b}\right)\sum_{n=1}^{N}\frac{1}{n^2}f(n)\sin\left(\frac{n\pi}{L}x\right)e^{-\frac{n\pi}{L}y}, \qquad (1.22)$$

where

$$f(n) = \sin\left(\frac{n\pi}{2}\right)\cos\left(\frac{n\pi b}{2L}\right). \qquad (1.23)$$

Now, the negative gradient of Eq. (1.22) gives the electric field distribution, as expressed in Eq. (1.24).

$$\vec{E}(x, y) = \frac{4}{\pi}\left(\frac{V_0}{L-b}\right)\sum_{n=1}^{N}\frac{1}{n}f(n)e^{\frac{-n\pi}{L}y}\left[-\cos\frac{n\pi x}{L}\hat{x} + \sin\frac{n\pi x}{L}\hat{y}\right]. \quad (1.24)$$

where $\hat{x}$ and $\hat{y}$ are the unit vectors in the x- and y-directions, respectively. For this nDEP system, the potential energy can then be written as Eq. (1.25).

$$U(x, y) = -\frac{24\varepsilon_0 \tilde{f}_{CM}(\omega)V_p}{\pi^2}\left(\frac{V_0}{L-b}\right)^2$$

$$\sum_{n=1}^{N}\sum_{m=1}^{N}\frac{1}{nm}f(n)f(m)\cos\left(\frac{(n-m)\pi}{L}x\right)e^{-\frac{(n+m)\pi}{L}y}. \quad (1.25)$$

By calculating the negative gradient of the potential energy shown in Eq. (1.25), the nDEP force applied to the particle is obtained.

$$\vec{F}_e(x, y) = -F_{mag}\sum_{n=1}^{N}\sum_{m=1}^{N}f(n)f(m)e^{-\frac{(n+m)\pi}{L}y}$$

$$\left[\left(\frac{n-m}{nm}\right)\sin\left((n-m)\pi\frac{x}{L}\right)\hat{x} + \left(\frac{n+m}{nm}\right)\cos\left((n-m)\pi\frac{x}{L}\right)\hat{y}\right], \quad (1.26)$$

where F_{mag} is defined by Eq. (1.27).

$$\vec{F}_{mag}(x, y) = \frac{24\varepsilon_0 \tilde{f}_{CM}(\omega)V_p}{\pi L}\left(\frac{V_0}{L-b}\right)^2. \quad (1.27)$$

Figure 1.8 illustrates an example of the discussed nDEP system. In Fig. 1.8a, a picture of the chip is shown. A microscopy image of the IDT design is shown in Fig. 1.8b. Also, Fig. 1.8c is a picture taken with a confocal microscope (side view). The microfluidic chip is used to produce an nDEP force on the particles, and based on the energy simulation results shown in Fig. 1.7b, the particles move upward. The red and blue dashed lines stand for the vertical positions of the levitated particles and a particle stuck on the chip surface, respectively. The particle on the chip surface is used as a reference to measure the levitation height of other particles. Also, a brightfield microscope image of the particles is shown in Fig. 1.8d, which then are levitated using the nDEP forces and thus are out of focus in Fig. 1.8e.

The plots in Fig. 1.9 illustrate the experimental results and theoretical predictions for the particle levitation based on the nDEP chip shown in Fig. 1.9. These results show the effect of the applied electrical potential and its frequency on the particle levitation height. The theoretical calculations are performed for various CM factors.

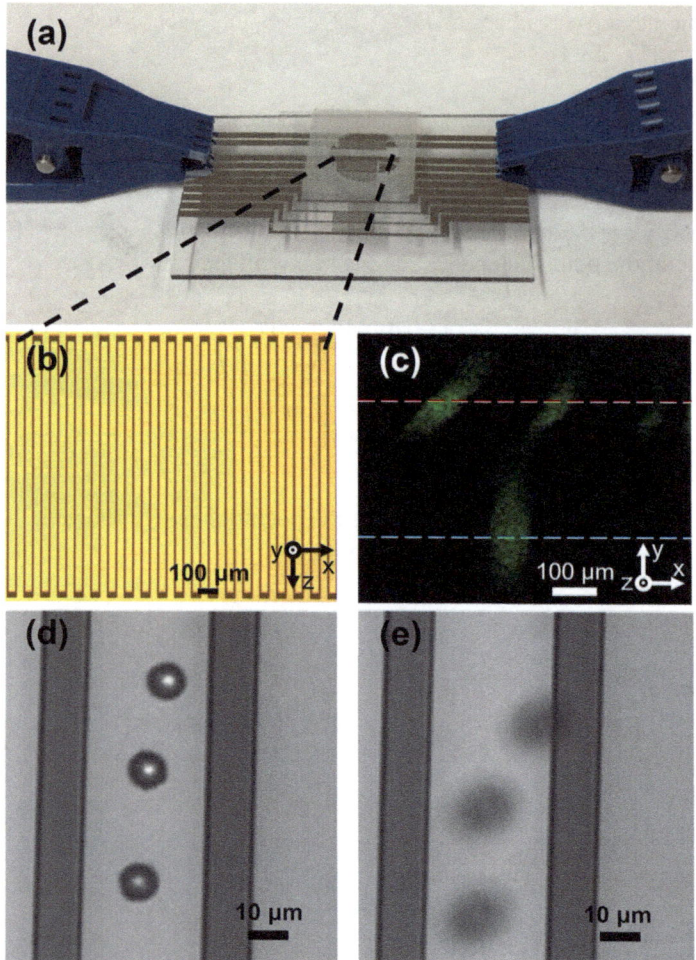

Fig. 1.8 nDEP Experimental setup. The chip and connectors are illustrated in (a) with a magnified view of the IDT in (b). A confocal microscopy image of the levitated particles above the surface is shown in (c). Representative optical microscopy images for the cases of (d) off and (e) on DEP forces are shown. Reprinted with permission from Springer Nature: Springer Nature, Microfluidics and Nanofluidics, Quantifying the dielectrophoretic force on colloidal particles in microfluidic devices, Abedini-Nassab et al., COPYRIGHT (2022) [121]

Thus, by calculating the CM factors based on these plots, one can predict the particle levitation in the proposed nDEP system.

In order to maximize the nDEP forces, we need to design a circuit that delivers most of the voltage to the fluid. Hence, we need to consider the effect of the chip and the connecting wires in an equivalent circuit model [138]. The electrode surface charge density and voltage are used to calculate the capacitance per unit length of the IDT array. The surface charge density can be calculated as Eq. (1.28) [121].

Fig. 1.9 Experimental measurement and theoretical prediction results for particle levitation are shown. The average vertical position with standard deviations is plotted. The applied frequencies are (a) 100 kHz, (b) 215 kHz, (c) 464 kHz, (d), 1 MHz, (e) 2.15 MHz, (f) 4.64 MHz, (g) 10 MHz, and (h) 20 MHz. Here, "Contact Force" means the force magnitude that the beads resting on the substrate experience. The theoretical curves show the particle positions for CM factors of −0.5 (solid line), −0.3 (dashed line), −0.1 (dashed-dotted line), and −0.05 (dotted line). Reprinted with permission from Springer Nature: Springer Nature, Microfluidics and Nanofluidics, Quantifying the dielectrophoretic force on colloidal particles in microfluidic devices, Abedini-Nassab et al., COPYRIGHT (2022) [121]

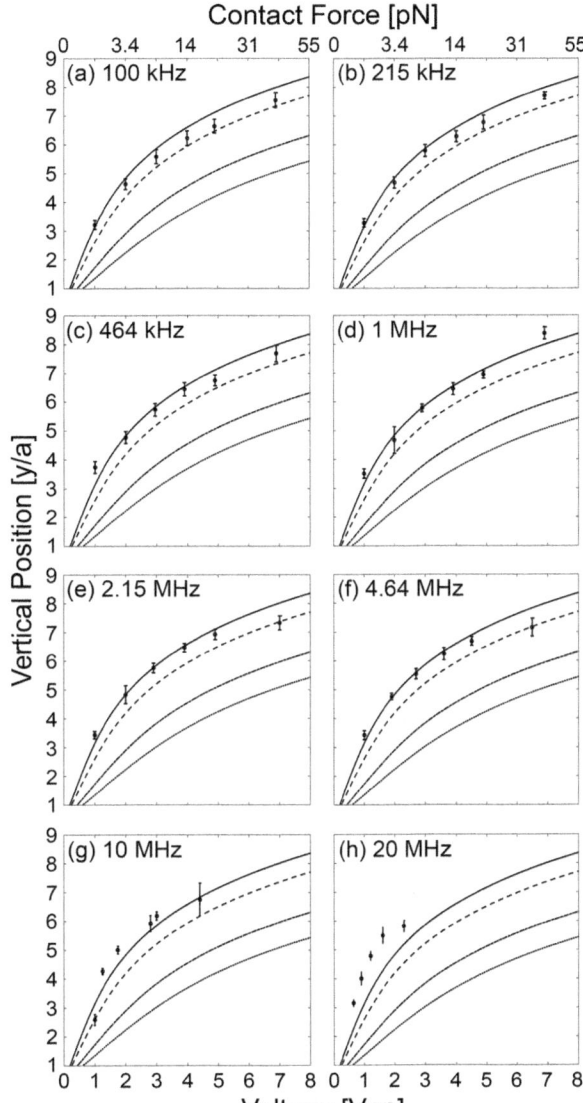

$$\delta(x) = \frac{4\varepsilon}{\pi}\left(\frac{V_0}{L-b}\right)\sum_{n=1}^{N}\frac{1}{n}f(n)\sin\frac{n\pi x}{L}. \qquad (1.28)$$

The capacitance is derived by integrating the total charge on the IDT surface Eq. (1.29) [121].

$$C_f = \left(\frac{2\varepsilon A}{L-b}\right) \sum_{n=1}^{N} \frac{f(n)}{(n\pi)^2} \cos\left(\frac{n\pi b}{2L}\right). \tag{1.29}$$

The electrical resistance of the IDT can be written as Eq. (1.30) [121].

$$R_f = \left[\left(\frac{2\sigma A}{L-b}\right) \sum_{n=1}^{N} \frac{f(n)}{(n\pi)^2} \cos\left(\frac{n\pi b}{2L}\right)\right]^{-1}, \tag{1.30}$$

where σ is the electric conductivity. The resistance and inductance of the wire leads and the power supply are also needed to be included in calculations. The inductance between a pair of wires is approximately written as

$$L_p = \frac{\mu_0}{\pi} \cosh^{-1}\left(\frac{S}{w}\right), \tag{1.31}$$

where μ_0, S, and w stand for the magnetic permeability of vacuum, the center-to-center distance between the wires, and their width, respectively. The resulting circuit model for the nDEP system is illustrated in Fig. 1.10a. The transfer function of this circuit is calculated as Eq. (1.32) [121].

$$\frac{\varphi_{out}}{\varphi_{in}} = \frac{\left(1 + R_t R_f^{-1} - \omega^2 \omega_R^{-2}\right) - j\omega\left(\omega_L^{-1} + \omega_C^{-1}\right)}{\left(1 + R_t R_f^{-1} - \omega^2 \omega_R^{-2}\right)^2 + \omega^2\left(\omega_L^{-1} + \omega_C^{-1}\right)^2}, \tag{1.32}$$

where $\omega_R = \sqrt{L_t C_f}^{-1}$, $\omega_L = R_f L_t^{-1}$, and $\omega_C = \left(R_t C_f\right)^{-1}$. The real part, imaginary part, and magnitude of Eq. (1.32) are plotted in Fig. 1.10b, for a sample IDT design. Based on these plots, the sample IDT design and experimental setup can appropriately operate up to tens of MHz. But, working at higher frequencies will need more complicated circuit designs.

Microfluidic chips in which DEP-based traps [139, 140] are included have been used to trap particles in a media and form particle arrays [141, 142], similar to the other array-based methods. Some size-selective trapping systems also have been introduced [143]. In addition to DEP, other electrokinetic phenomena (e.g., AC electro-osmosis (ACEO) [144] and AC electrothermal effects (ACET) [145]) will also move the particles in a fluid. This consideration is particularly important in fluids of high ionic strength, where more complicated calculations are needed. Several review papers discussing these phenomena are available for more information [145–149].

1.1.2.3   Digital Microfluidics

A well-established liquid handling technique, called digital microfluidics, is also used to manipulate droplets of liquids, in which particles may exist, in microfluidic chips.

Fig. 1.10 Equivalent circuit model for the IDT design. (a) The circuit model is shown, where R_t and L_t stand for the resistance and inductance of the wire leads, respectively. Also, R_f and C_f stand for the resistance and capacitance of the interdigitated electrodes and the media, respectively. (b) The real part (dotted line), imaginary part (dashed line), and the magnitude of the equivalent circuit as a function of the applied frequency for a sample IDT design are plotted. Reprinted with permission from Springer Nature: Springer Nature, Microfluidics and Nanofluidics, Quantifying the dielectrophoretic force on colloidal particles in microfluidic devices, Abedini-Nassab et al., COPYRIGHT (2022) [121]

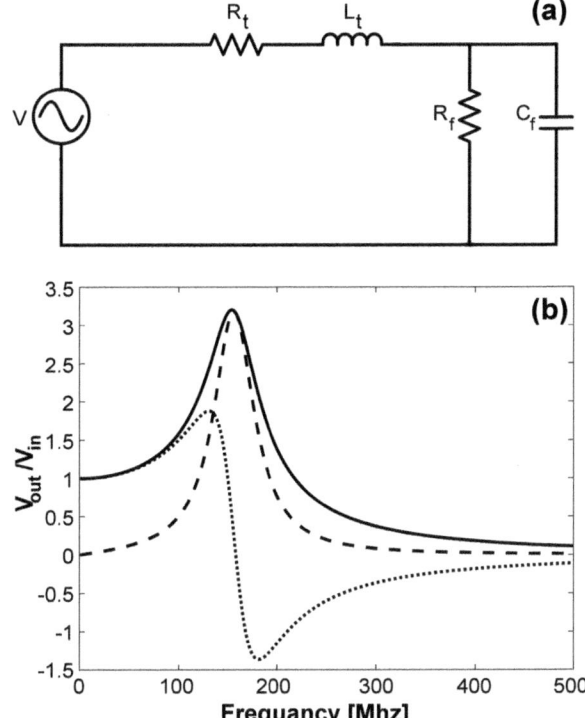

Digital microfluidics works based on manipulating droplets on an array of electrodes using electric forces. The electrodes are covered by a hydrophobic insulator, and tiny droplets on them are independently addressed to be moved, by applying excitation electric signals to the right electrodes. By moving the droplets, many different droplet manipulating tasks, such as merging, mixing, and splitting, are made possible.

The digital microfluidic devices can be based on a single open plate or a plate pair (*i.e.*, closed). In the single plate format, the substrate consists of actuator electrodes carrying the manipulation signal and the ground electrodes. In the closed format, the plate pair sandwiches the droplets and consists of a substrate plate with an array of electrodes on it, carrying the manipulation signal, and a top electrically grounded conductive indium tin oxide (ITO) plate.

The digital microfluidics operation initially was explained based on the electrowetting principle, where the contact angle of the droplets on the chip surface is manipulated by the applied voltage and is described by Eq. (1.33), known as the Young-Lippmann equation [150].

$$\cos\theta(\varphi) = \cos\theta(0) + \frac{1}{2}\frac{\varepsilon_0\varepsilon}{d\gamma}\varphi^2, \tag{1.33}$$

Fig. 1.11 A sample droplet transport in a digital microfluidic chip is shown. Frames at various time points are illustrated in panels from left to the right. The figure is taken from [152] with permission, Copyright 2020 MDPI, under the terms and conditions of the Creative Commons Attribution (CC BY) license (http://creativecommons.org/licenses/by/4.0/)

where d, γ, $\theta(0)$, and $\theta(\varphi)$ are the dielectric thickness, the liquid-gas surface tension, the contact angle before excitation, and the contact angle after excitation, respectively. More recently, by considering the capacitances of each element in the system and deriving the total capacitance, the stored energy is calculated to be [151]

$$U(f,x) = \frac{L}{2}\left(x\sum_i \frac{\varepsilon_0\varepsilon_{drop,i}\varphi_{drop,i}^2(j2\pi f)}{d_i} + (L-x)\sum_i \frac{\varepsilon_0\varepsilon_{m,i}\varphi_{m,i}^2(j2\pi f)}{d_i}\right),$$

(1.34)

where f, $\varepsilon_{drop,i}$, $\varphi_{drop,i}$, $\varepsilon_{m,i}$, and $\varphi_{m,i}$ stand for the applied frequency, the relative permittivity of the droplet, the voltage drop on the droplet, the relative permittivity of the surrounding media, and the voltage drop on the surrounding media, respectively. d_i stands for the thickness of layer i (e.g., the dielectric layer, the top and bottom hydrophobic layers, the droplet, and the surrounding media). Now, the applied force can be calculated by taking the derivative of Eq. (1.35) [151].

$$F(f) = \frac{\partial U(f,x)}{\partial x} = \frac{L}{2}\left(\sum_i \frac{\varepsilon_0\varepsilon_{drop,i}\varphi_{drop,i}^2(j2\pi f)}{d_i} - \sum_i \frac{\varepsilon_0\varepsilon_{m,i}\varphi_{m,i}^2(j2\pi f)}{d_i}\right).$$

(1.35)

A sample droplet movement is illustrated in Fig. 1.11. In this figure, the droplet position at five different time points is shown. The droplet is maintained between electrodes, by exciting individual electrodes, sequentially.

Although digital microfluidics tools offer precise microdroplet manipulation, usually voltages of ~50 V or more is needed for appropriate operation. These relatively high voltages can be challenging. Also, increasing the electrode numbers would result in a complex wiring system. Thus, to keep the chips simple, the number of droplets or particles in these methods is typically limited.

1.1.2.4 Other Particle Manipulation Methods

Other similar methods, based on other forces, for manipulating particles are introduced. For example, optical tweezers have also been used to sort particles and cells.

In this approach, optical waves provide the required energy distribution to move the particles to specific spots. Precise particle manipulation is offered with this technique; however, delivering heat to the cells and cell photo-damage problems, affecting the cell metabolism, growth, and division, are considered a concern [88, 89].

Microcoil arrays are proposed to manipulate magnetically labeled particles [101] too. These systems operate similarly to the ones based on electrowetting; however, as opposed to the electric forces, magnetic forces are used. In both methods, the trajectory of single particles and cells can precisely be defined. But, a large number of electrodes or microcoils are required, which may end up in a complicated wiring system. Also, in these systems, the number of single particles and cells to be manipulated is usually limited (*i.e.*, less than a few hundred cells).

In some other works, a simpler system based on magnetic thin films magnetized in an external magnetic field is proposed [102–104]. This approach eliminates the need for microcoil arrays or complicated wiring systems. But, since all the particles respond to the external magnetic field, control of individual particles, which is possible in coil-based systems, is not offered. We will discuss this concept in more detail in the next chapters.

All SCA tools have their own advantages and disadvantages. But, to perform better single-cell analysis with more remarkable results, SCA tools with a high level of scalability, flexibility, and automation, capable of capturing rare but important dynamic events within the heterogeneous cell populations, are needed. That's where magnetophoretic circuits, based on a novel magnetic manipulation technique, can play an important role. To provide a bigger picture, I first provide a more general overview of magnetic methods used in biology and medicine, and then, I will discuss the magnetic theory required for understanding the concepts of magnetophoretic circuits.

1.2 Applications of Magnetic Techniques in Biology

Although our main goal in this text is to study the magnetophoretic circuits, here I provide a bigger picture of the magnetic methods used in medicine. Since magnetic materials respond to external magnetic fields, they have found numerous applications in different fields, including medicine and biology. The magnetic fields pass through tissues and organs and can provide magnetic forces on available magnetic materials, remotely. Thus, inside the body, objects can be controlled noninvasively (*i.e.*, without requiring wires or tubes inside the body). Because of these capabilities, many researchers are interested in magnetic manipulation techniques.

1.2.1   In Vivo Applications

In vivo studies are the ones performed inside the body of a living organism, animal, or plant. Here, we consider some popular magnetic-based techniques with in vivo applications.

1.2.1.1   Magnetic Imaging

Magnetic-based medical imaging is one of the popular applications of magnetic materials. In these methods, magnetic radiation helps in providing pictures of the organ or the processes in vivo. A noninvasive well-established method is called magnetic resonance imaging (MRI), which works based on the nucleic magnetic resonance (NMR) phenomenon. Professor Isidor I. Rabi, the winner of the Nobel Prize, first observed NMR, where atomic nuclei in a sufficiently strong magnetic field absorb or re-emit electromagnetic radiation. The characteristics of the resulting wave depend on the target atom. Thus, different tissues with different atom compositions create different signals and so can be identified. Based on these signals, the spatial distribution of various properties in the organs is then visualized. But the contrast based on the intrinsic water content in the tissues is normally limited. Hence, contrast agents are used to shorten the relaxation time (e.g., gadolinium ion and iron oxide particles).

MRI recognizes and characterizes numerous anatomical and physiological parameters. For example, it can identify the tissue cellular phagocytic activity, stiffness, and level of renal filtration [153]. Moreover, the MRI technique can be used with the help of superparamagnetic nanoparticles to track cells in vivo [154–156]. Some other examples of the medical imaging applications of MRI are the identification of active infections and inflammations [157], clinical detections [158], tumor imaging [159], and classification of cancer tumors [160]. Additionally, the combination of MRI with other techniques has resulted in good outcomes to answer medical needs. For example, together with clinical diagnostic methods, it has been used in the prediction of conversion from mild cognitive impairment to Alzheimer's disease [161].

Magnetic Particle Imaging (MPI) is an alternative relatively new magnetic imaging method, in which nonlinear magnetization of ferromagnetic nanoparticles is used for imaging purposes [162]. In this technique, after adding magnetic nanoparticles to the sample, it is exposed to a strong DC magnetic field superimposed on a weak field modulation. The purpose of applying the strong DC magnetic field to the magnetic nanoparticles is to provide a flat magnetization curve, which cancels the harmonic signals. Now, applying this strong magnetic field to the entire sample, except for a tiny "field-free" area, leaves only signals coming from that field-free area [163]. Now, by moving the sample, the magnitudes of the harmonics from the entire sample are measured [162]. This technique is an appropriate method for fast in vivo three-dimensional dynamic imaging [164], with applications in molecular imaging and stem cell monitoring [165], clinical diagnostics, and image-guided intervention [166].

1.2.1.2 Therapeutic Methods

The capabilities of magnetic nanoparticles in therapeutic applications are widely accepted in the field. For example, they are used in magnetically guided drug-targeting methods. Delivering the drug directly to the target site not only enhances the therapeutic efficiency but also reduces the side effects of the drugs on healthy tissues. In this technique, the magnetic nanoparticles or microparticles carry the drug of interest and in an appropriate externally applied magnetic field gradient move toward the desired spot(s) in the body [167, 168]. For example, cancer drugs can be magnetically guided to targeted tumors [168, 169]. In an alternative approach, as opposed to magnetically carrying the anti-cancer drugs to the tumors, novel magnetic anti-cancer compounds are used. In this method, the magnetic materials have two roles: (i) respond to the external field and move towards the target, and (ii) show anti-cancer characteristics [170]. Currently, these methods are mostly in animal studies; however, some clinical experiments are already performed too [171].

Another example of therapeutic applications of magnetic materials is magnetic fluid hyperthermia for treating cancer [172] and other diseases [173, 174]. In this method, oscillating magnetic fields are applied to magnetic fluids (nanoparticles) [175] to heat the target tissue to 43–46 °C [176] and kill the unhealthy cells. Furthermore, in some studies, magnetically produced heat is used for releasing the drug in the delivery systems [177, 178]. In this method, first, the drug-carrying magnetic particles are moved to the target spot. Then, an oscillating magnetic field is applied to produce the required heat for releasing the drug(s).

1.2.2 In Vitro Applications

In addition to remotely manipulating the magnetic particles inside the body, the magnetic forces are used for sample preparations (e.g., purification and separation) biotests (e.g., detections and drug screening), and other in-lab experiments. Here, some examples of these out-of-body experiments, called in vitro experiments, are provided.

1.2.2.1 Purification and Separation

In most bio-labs, a routine task is to separate biological particles (e.g., DNA, RNA, and proteins) using magnetic forces. Affinity-coated magnetic particles bind to the target molecule and transport them in an externally applied magnetic field gradient.

Commercial magnetic columns are widely used for separation and purification purposes.

A good example of magnetic bio-separation is isolating nucleic acids from the other cellular contents of the lysed cells for genomic studies. In these experiments, typically magnetic particles are coated with oligonucleotides complementary to a section of the target molecule. When introduced to the sample, they capture the molecule of interest, and the complex is magnetically isolated (*i.e.*, in an external magnetic field gradient). To target messenger ribonucleic acids (mRNA) from a sample, its poly A tail can be targeted. Commercially available magnetic beads with Oligo $(dT)_{25}$ coating (Dynabeads®) can do the job [179]. But when the mRNAs are cleaved, other mRNA sections are targeted. In the case of purifying deoxyribonucleic acid (DNA) molecules, oligonucleotide functionalized magnetic particles, amino-modified silica particles [180], or carboxyl group-containing magnetic particles [181, 182] are used.

Successful purification of recombinant proteins using techniques based on affinity interactions between the Histidine (His) tag of the target proteins and metal ions is possible [183]. Magnetic beads covered with Ni-nitriloacetic acid ligands are used to isolate 6xHis-Tagged proteins from the samples [184]. Similar methods for purifying different proteins based on the immobilization of affinity ligands on magnetic particles are proposed [185].

Another important example of using magnetic materials in vitro is magnetic cell manipulation. By magnetically labeling live cells (*i.e.*, conjugation to antibody-labeled magnetic nanoparticles or beads or having cells uptake magnetic nanoparticles), they respond to the magnetic field gradients. For example, specific cell surface antigens are targeted with specific antibodies on magnetic particles to link. Then, the labeled cells can be detected or purified [186]. This method is a well-established technique in cell separation [187–190].

1.2.2.2 Bio-Sensing

Asynchronous magnetic bead rotation (AMBR) is a magnetic sensing method with many biological applications, including label-free DNA analysis [191], single bacterial cell detection [192], monitoring bacterial growth and their drug sensitivity [193–195], and detecting proteins [196]. In this method, superparamagnetic particles exposed to a low-frequency external rotating magnetic field follow it synchronously. However, at frequencies higher than a critical frequency, the particles fail to chase the external magnetic field. Since the particle rotational speed is proportional to their effective volume, after bounding to another particle, such as any biomolecule, they move at a different speed [195]. Thus, the speed change can be used as a presence or growth detection parameter.

Giant magnetoresistance (GMR) is a spin-dependent electron scattering phenomenon, based on which magnetic detection is possible. GMR sensors are metallic multilayer structures composed of alternating ferromagnetic and non-magnetic thin films. In an externally applied magnetic field, the directions of the magnetization of the two adjacent magnetic layers are either parallel or antiparallel. The change in the magnetization direction of these layers (*i.e.*, antiparallel to parallel or wise versa) results in a significant change in the electrical resistivity [197, 198]. This change in electrical resistivity is due to the conservation of electron spin over the thickness of the conductive layer. In the case of antiparallel alignment, electrons moving through the multilayer structure face electrons with opposite spin and get scattered [198]. To use this structure as a sensor, one needs to change the magnetic field near the sensor. Thus, if any biological particle carrying magnetic nanoparticles or microparticles moves close to the sensor, it affects the applied magnetic field and changes the resistivity of the structure. To make this sensor sensitive to specific antigens, it is possible to coat the sensor surface with an antibody specific to that target antigen to keep it close to the sensor.

A good example of the applications of GMR sensors is a wash-free magnetic bioassay for detecting the Influenza A virus in swine nasal swab samples [199]. After mixing the biotinylated Influenza A virus detection antibody with magnetic particles and the biological sample, the target protein is captured and then placed on the GMR chip. The antibody on the chip surface captures the target analyte-detection antibody-magnetic bead complex. The resulting sandwich structure produces the detection signal.

GMR biosensors have also been used to detect ovarian cancer biomarkers [199]. Figure 1.12 illustrates a picture of the chip. The platform detects numerous protein biomarkers of human diseases. An array of GMR is created, while each sensor has an antibody sensitive to a biomarker of interest. Figure 1.13 shows the capture process in four steps. Using Ademtech 200 nm magnetic beads, each composed of ~1000 magnetic nanoparticles with an average magnetic moment of ~2.3 × 10–16 emu, the researchers have detected cancer antigen 125 (CA125 II), human epididymis protein 4 (HE4), and interleukin 6 (IL6), with detection limits of 3.7 U/mL, 7.4 pg/mL, and 7.4 pg/mL, respectively. Similar methods are used for the early detection of multiple cirrhosis biomarkers [200], where intercellular adhesion molecule-1 (sICAM-1) and mac-2 binding protein glycosylation isomer (M2BPGi) are identified.

A similar multilayer structure, called the spin-valve sensor, by adding an antiferromagnetic (pinning) layer to enhance the sensitivity of GMR sensors is also engineered. Here, two steps are employed. First, the temperature is increased to values higher than the knee temperature, and the pinned layer coupling is deactivated. Then, by cooling the system and applying a fixed magnetic field to define its magnetization direction, the pinned direction is fixed [201–203].

Tunneling magnetoresistive (TMR) sensors have also been used in bio-applications. In this technique, a thin insulating layer is sandwiched with the ferromagnetic layers, and a tunneling current passes through the insulating layer. In TMR biosensors, the change in resistance due to the stray magnetic field from the magnetic particles is monitored, and thus, it can quantitatively determine the captured biomarker contents [204].

Researchers have proposed a rapid Escherichia coli bacteria detection tool based on a TMR sensor [205]. In this method, the target is labeled by magnetic particles producing a magnetic fringe field in an external field. This signal is detected by the

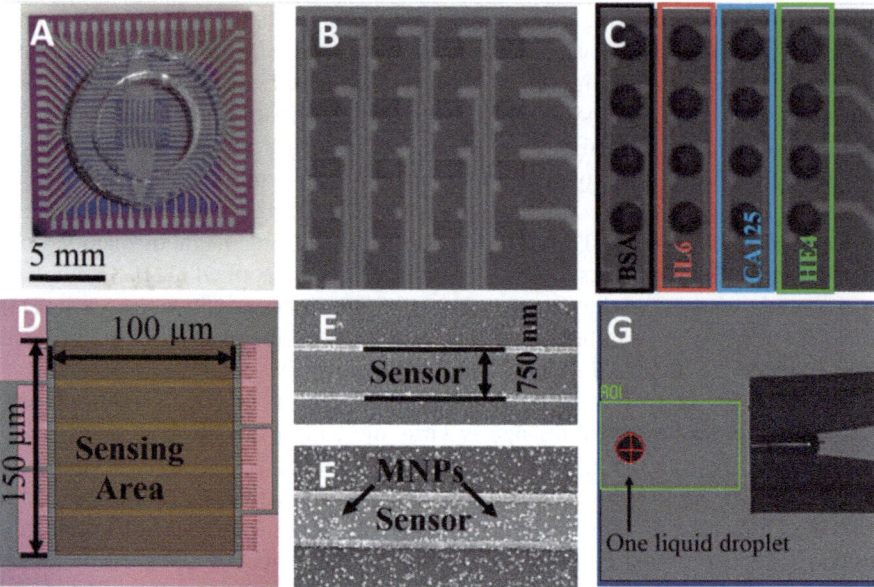

Fig. 1.12 (A) An example GMR array chip is illustrated. The sensor array (B) before and (C) after introducing the samples is shown. (D) A close view of the sensors is shown. Scanning electron microscopy images of a GMR sensor strip (E) before and (F) after magnetic nanoparticles bound to its surface are shown. (G) A piezo device distributes the nanoparticles. Reprinted with permission from T. Klein et al., 2019. Biosensors and Bioelectronics 126, 301–307 [199]. Copyright 2020 Elsevier

TMR sensor. Detection of ricin using magnetic nanoparticles and TMR sensors is also reported [206]. By combining a magnetic immuno-chromatographic test strip and a TMR sensor, they have been able to overcome, as they claim, the optical signal detection challenges in some traditional biosensors.

Detection of pathogens in food based on TMR sensors is also reported [207]. Genomic DNA extracted from the pathogenic bacterium Listeria monocytogenes with a sensitivity below the nM range is detected. In this method, the sample is hybridized with complementary target DNA. Then, streptavidin-coated magnetic nanoparticles with a diameter of ~250 nm are introduced. An external magnetic field is applied to the chip, nanoparticles interact with the biotinylated DNA on the sensor surface, and the sensor output signal is recorded.

Magnetorelaxometry (MRX) is another magnetic-based detection method. In the MRX-based sensors, first, a magnetic field is applied to the magnetic particles. The relaxation behavior of the magnetic particles after switching off the magnetic field reveals their magnetic properties. The dynamics of the magnetic nanoparticle magnetic moments are described using Brownian and the Néel relaxations [208, 209]. The Brownian relaxation time constant is expressed as

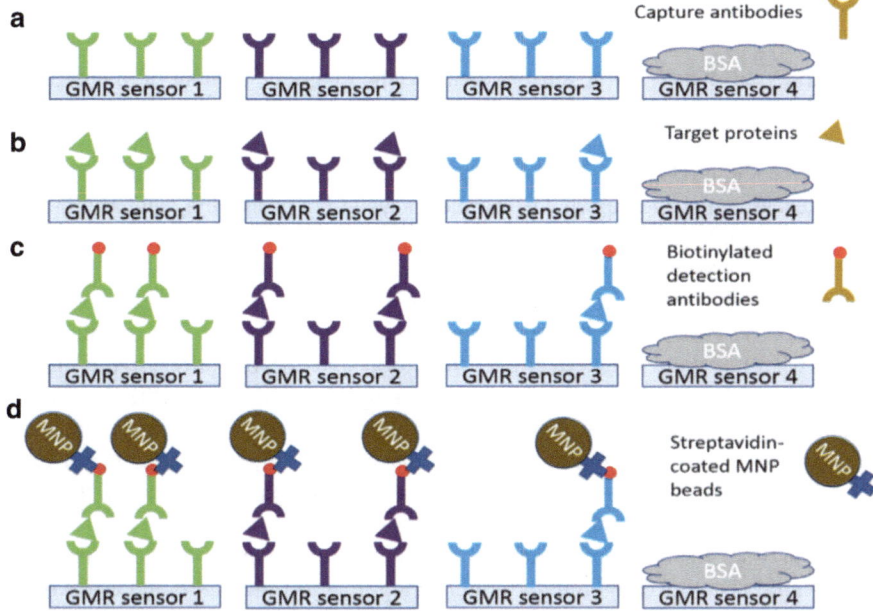

Fig. 1.13 A schematic of a GMR assay for the detection of multiple proteins is shown. (A) Four GMR sensors with different capture antibodies. (B) Samples with target proteins are introduced. (C) Biotinylated detection antibodies are added to selectively bind to the specific target analyte. (D) Streptavidin-coated magnetic nanoparticles are added, and then GMR signals are booked. Reprinted with permission from T. Klein et al., 2019. Biosensors and Bioelectronics 126, 301–307 [199]. Copyright 2020 Elsevier

$$\tau_B = \frac{3\eta V_h}{k_B T}, \tag{1.36}$$

where Vh, k_B, and T are the particle hydrodynamic volume, the Boltzmann constant, and the temperature, respectively. The zero-field Néel relaxation time constant is expressed by

$$\tau_N = \tau_0 e^{\frac{K_u V_p}{k_B T}}, \tag{1.37}$$

where τ_0, K_u, and V_p depict the damping time, effective magnetic anisotropy constant, and particle volume, respectively. By defining the effective relaxation time as $\tau = \frac{\tau_B \tau_N}{\tau_B + \tau_N}$, the time-dependent net magnetic flux density will result as expressed in Eq. (1.38).

$$B(t) = B_0 e^{-t/\tau} + B_{Offset}, \tag{1.38}$$

where B_0 and B_{Offset} are the magnetic field flux at relaxation time and the offset, respectively.

In bio-experiments, MRX sensors measure the magnetic content of magnetically labeled bioparticles based on different methods. One sensitive sensor to be combined with MRS with numerous bio-applications is the superconducting quantum interference device (SQUID) [208–214]. Hall sensors are also used in other magnetometer chips. Researchers have reported using such a system to profile magnetically labeled single cancer cells [215].

The last magnetic sensor with applications in biology to be reviewed here is the one based on the anisotropic magnetoresistive (AMR) effect. This effect is based on the electric resistivity change of some magnetic materials as a result of the change between their magnetization orientation and the direction of the electric current passing through them [216]. As an example, a disposable AMR-based chip is introduced to detect DNA labeled with magnetic particles [217]. In this technique, researchers hybridized the magnetically labeled single-stranded target DNA with a DNA probe, which resulted in a relatively linear sensor response in the DNA concentration range of 4.5 to 18 pmol.

To conclude, various particle and droplet manipulation methods were discussed. These methods are grouped into two main categories of flow-based and array-based systems. In flow-based systems, the particles are analyzed and/or sorted while being transported in a fluid flow. FACS and droplet-based microfluidic chips are two important examples, each of which has advanced the field. In the array-based method, the particles are assembled into an array on the chip ready to be analyzed in parallel. To transport the particles, various forces, including hydrodynamic, acoustic, magnetic, and electric forces, are used. Since magnetomicrofluidic circuits operate based on magnetic forces, this force was discussed in more detail. Examples of using magnetic techniques in biology including in vitro and in vivo applications were provided.

References

1. Gijs, M. A., Lacharme, F., & Lehmann, U. (2010). Microfluidic applications of magnetic particles for biological analysis and catalysis. *Chemical Reviews, 110*(3), 1518–1563.
2. Abedini-Nassab, R., & Eslamian, M. (2014). Recent patents and advances on applications of magnetic nanoparticles and thin films in cell manipulation. *Recent Patents on Nanotechnology, 8*(3), 157–164.
3. Lin, Z., et al. (2020). Rapid assessment of surface markers on cancer cells using immunomagnetic separation and multi-frequency impedance cytometry for targeted therapy. *Scientific Reports, 10*(1), 3015.
4. Hsiao, Y.-C., et al. (2019). Capturing magnetic bead-based arrays using perpendicular magnetic anisotropy. *Applied Physics Letters, 115*(8), 082402.
5. Yu, W., et al. (2020). A ferrobotic system for automated microfluidic logistics. *Science Robotics, 5*(39), eaba4411.
6. Saliba, A. E., et al. (2010). Microfluidic sorting and multimodal typing of cancer cells in self-assembled magnetic arrays. *Proceedings of the National Academy of Sciences, 107*(33), 14524–14529.

7. Cha, J., & Lee, I. (2020). Single-cell network biology for resolving cellular heterogeneity in human diseases. *Experimental Molecular Medicine, 52*(11), 1798–1808.
8. de Armas, L. R., et al. (2019). Single cell profiling reveals PTEN overexpression in influenza-specific B cells in aging HIV-infected individuals on anti-retroviral therapy. *Scientific Reports, 9*(1), 2482.
9. Gantner, P., et al. (2020). Single-cell TCR sequencing reveals phenotypically diverse clonally expanded cells harboring inducible HIV proviruses during ART. *Natural Communication, 11*(1), 4089.
10. Chen, H., et al. (2020). Single cell transcriptome revealed SARS-CoV-2 entry genes enriched in colon tissues and associated with coronavirus infection and cytokine production. *Signal Transduction and Targeted Therapy, 5*(1), 121.
11. Malone, A. F., & Humphreys, B. D. (2019). Single-cell transcriptomics and solid organ transplantation. *Transplantation, 103*(9), 1776–1782.
12. Pei, H., et al. (2020). Recent advances in microfluidic technologies for circulating tumor cells: Enrichment, single-cell analysis, and liquid biopsy for clinical applications. *Lab on a Chip, 20*(21), 3854–3875.
13. Su, X., et al. (2021). Clonal evolution in liver cancer at single-cell and single-variant resolution. *Journal of Hematology & Oncology, 14*(1), 22.
14. Lawson, D. A., et al. (2018). Tumour heterogeneity and metastasis at single-cell resolution. *Nature Cell Biology, 20*(12), 1349–1360.
15. Saadatpour, A., et al. (2015). Single-cell analysis in cancer genomics. *Trends in Genetics, 31*(10), 576–586.
16. Dhar, M., et al. (2015). Research highlights: Microfluidic-enabled single-cell epigenetics. *Lab on a Chip, 15*(21), 4109–4113.
17. Rosenfeld, N., et al. (2005). Gene regulation at the single-cell level. *Science, 307*(5717), 1962–1965.
18. Cai, L., Friedman, N., & Xie, X. S. (2006). Stochastic protein expression in individual cells at the single molecule level. *Nature, 440*(7082), 358–362.
19. Bjorklund, A. K., et al. (2016). The heterogeneity of human CD127(+) innate lymphoid cells revealed by single-cell RNA sequencing. *Nature Immunology, 17*(4), 451–460.
20. Churchill, M. J., et al. (2016). HIV reservoirs: What, where and how to target them. *Nature Reviews Microbiology, 14*(1), 55–60.
21. Eriksson, S., et al. (2013). Comparative analysis of measures of viral reservoirs in HIV-1 eradication studies. *PLoS Pathogens, 9*(2), e1003174.
22. Ho, Y. C., et al. (2013). Replication-competent noninduced proviruses in the latent reservoir increase barrier to HIV-1 cure. *Cell, 155*(3), 540–551.
23. Lubetzky, M. L., et al. (2021). Urinary Cell mRNA profiles predictive of human kidney Allograft status. *Clinical Journal of the American Society of Nephrology, 16*(10), 1565–1577.
24. Muthukumar, T., et al. (2021). Single Cell RNA-sequencing of urinary cells and defining the immune landscape of rejection in human kidney Allografts. In *2021 American Transplant Congress*. Wiley.
25. Wang, D., & Bodovitz, S. (2010). Single cell analysis: The new frontier in 'omics.' *Trends in Biotechnology, 28*(6), 281–290.
26. Reya, T., et al. (2001). Stem cells, cancer, and cancer stem cells. *Nature, 414*(6859), 105–111.
27. Chen, K., Huang, Y. H., & Chen, J. L. (2013). Understanding and targeting cancer stem cells: Therapeutic implications and challenges. *Acta Pharmacologica Sinica, 34*(6), 732–740.
28. Shackleton, M., et al. (2009). Heterogeneity in cancer: Cancer stem cells versus clonal evolution. *Cell, 138*(5), 822–829.
29. Greaves, M., & Maley, C. C. (2012). Clonal evolution in cancer. *Nature, 481*(7381), 306–313.
30. Nowell, P. C. (1976). The clonal evolution of tumor cell populations. *Science, 194*(4260), 23–28.
31. Wills, Q. F., & Mead, A. J. (2015). Application of single-cell genomics in cancer: Promise and challenges. *Human Molecular Genetics, 24*(R1), R74–84.

32. Abedini-Nassab, R., Emami, S. M., & Nowghabi, A. N. (2021). Nanotechnology and acoustics in medicine and biology. *Recent Patents on Nanotechnology, 16*(3), 198–206.

33. Drafts, B. (2001). Acoustic wave technology sensors. *IEEE Transactions on Microwave Theory and Techniques, 49*(4), 795–802.

34. Liu, Y., et al. (2020). Materials, design, and characteristics of bulk acoustic wave resonator: A review. *Micromachines, 11*(7), 630.

35. Shi, J., et al. (2009). Continuous particle separation in a microfluidic channel via standing surface acoustic waves (SSAW). *Lab on a Chip, 9*(23), 3354–3359.

36. Barmatz, M., et al. (1985). *Acoustic particle separation.* U.S. Patent 4,523,682.

37. Ota, N., et al. (2019). Enhancement in acoustic focusing of micro and nanoparticles by thinning a microfluidic device. *Royal Society Open Science, 6*(2), 181776.

38. Guldiken, R., et al. (2012). Sheathless size-based acoustic particle separation. *Sensors, 12*(1), 905–922.

39. McKinnon, K. M. (2018). Flow cytometry: An overview. *Current Protocols in Immunology, 120*(1), 5.1.1–5.1.11.

40. Henry, T. C., & Brynildsen, M. P. (2016). Development of persister-facseq: A method to massively parallelize quantification of persister physiology and its heterogeneity. *Scientific Reports, 6*, 25100.

41. Bourseau-Guilmain, E., et al. (2016). Hypoxia regulates global membrane protein endocytosis through caveolin-1 in cancer cells. *Nature Communications, 7*, 11371.

42. Wolf, N. S., et al. (1993). In vivo and in vitro characterization of long-term repopulating primitive hematopoietic cells isolated by sequential Hoechst 33342-rhodamine 123 FACS selection. *Experimental Hematology, 21*(5), 614–622.

43. Prussin, C., & Metcalfe, D. D. (1995). Detection of intracytoplasmic cytokine using flow cytometry and directly conjugated anti-cytokine antibodies. *Journal of Immunological Methods, 188*(1), 117–128.

44. Scheffold, A., et al. (2000). High-sensitivity immunofluorescence for detection of the pro- and anti-inflammatory cytokines gamma interferon and interleukin-10 on the surface of cytokine-secreting cells. *Nature Medicine, 6*(1), 107–110.

45. Dewandre, A., et al. (2020). Microfluidic droplet generation based on non-embedded co-flow-focusing using 3D printed nozzle. *Scientific Reports, 10*(1), 21616.

46. Shang, L., Cheng, Y., & Zhao, Y. (2017). Emerging droplet microfluidics. *Chemical Reviews, 117*(12), 7964–8040.

47. Liu, Z., et al. (2020). Microfluidics for production of particles: Mechanism, methodology, and applications. *Small (Weinheim an der Bergstrasse, Germany), 16*(9), e1904673.

48. Pessi, J., et al. (2014). Microfluidics-assisted engineering of polymeric microcapsules with high encapsulation efficiency for protein drug delivery. *International Journal of Pharmaceutics, 472*(1–2), 82–87.

49. Wang, Y., et al. (2020). Advances of droplet-based microfluidics in drug discovery. *Expert Opinion on Drug Discovery, 15*(8), 969–979.

50. Park, J. W., et al. (2015). Live cell imaging compatible immobilization of Chlamydomonas reinhardtii in microfluidic platform for biodiesel research. *Biotechnology Bioengineering, 112*(3), 494–501.

51. Hajji, I., et al. (2020). Droplet microfluidic platform for fast and continuous-flow RT-qPCR analysis devoted to cancer diagnosis application. *Sensors and Actuators B: Chemical, 303*, 127171.

52. Abedini-Nassab, R., Pouryosef Miandoab, M., & Şaşmaz, M. (2021). Microfluidic synthesis, control, and sensing of magnetic nanoparticles: A review. *Micromachines, 12*(7), 768.

53. Yao, J., et al. (2019). The effect of oil viscosity on droplet generation rate and droplet size in a t-junction microfluidic droplet generator. *Micromachines, 10*(12), 808.

54. Ushikubo, F. Y., et al. (2014). Y- and T-junction microfluidic devices: Effect of fluids and interface properties and operating conditions. *Microfluidics and Nanofluidics, 17*(4), 711–720.

55. Huang, D., et al. (2020). Precise control for the size of droplet in T-junction microfluidic based on iterative learning method. *Journal of the Franklin Institute, 357*(9), 5302–5316.

56. Garstecki, P., et al. (2006). Formation of droplets and bubbles in a microfluidic T-junction—Scaling and mechanism of break-up. *Lab on a Chip, 6*(3), 437–446.

57. Zhang, J., et al. (2021). Microfluidic droplet formation in co-flow devices fabricated by micro 3D printing. *Journal of Food Engineering, 290*, 110212.

58. Cramer, C., Fischer, P., & Windhab, E. J. (2004). Drop formation in a co-flowing ambient fluid. *Chemical Engineering Science, 59*(15), 3045–3058.

59. Yin, Z., et al. (2020). Droplet generation in a flow-focusing microfluidic device with external mechanical vibration. *Micromachines, 11*(8), 743.

60. Lashkaripour, A., et al. (2021). Machine learning enables design automation of microfluidic flow-focusing droplet generation. *Nature Communications, 12*(1), 25.

61. Yaghmur, A., et al. (2019). A hydrodynamic flow focusing microfluidic device for the continuous production of hexosomes based on docosahexaenoic acid monoglyceride. *Physical Chemistry Chemical Physics, 21*(24), 13005–13013.

62. Rotem, A., et al. (2015). High-throughput single-cell labeling (Hi-SCL) for RNA-Seq using drop-based microfluidics. *PLoS ONE, 10*(5), e0116328.

63. Spencer, S. J., et al. (2016). Massively parallel sequencing of single cells by epicPCR links functional genes with phylogenetic markers. *The ISME Journal, 10*(2), 427–436.

64. Macosko, E. Z., et al. (2015). Highly parallel genome-wide expression profiling of individual cells using nanoliter droplets. *Cell, 161*(5), 1202–1214.

65. Klein, A. M., et al. (2015). Droplet barcoding for single-cell transcriptomics applied to embryonic stem cells. *Cell, 161*(5), 1187–1201.

66. Chokkalingam, V., et al. (2013). Probing cellular heterogeneity in cytokine-secreting immune cells using droplet-based microfluidics. *Lab on a Chip, 13*(24), 4740–4744.

67. Konry, T., Golberg, A., & Yarmush, M. (2013). Live single cell functional phenotyping in droplet nano-liter reactors. *Science Reports, 3*, 3179.

68. Konry, T., et al. (2011). Droplet-based microfluidic platforms for single T cell secretion analysis of IL-10 cytokine. *Biosensors and Bioelectronics, 26*(5), 2707–2710.

69. Leng, X., et al. (2010). Agarose droplet microfluidics for highly parallel and efficient single molecule emulsion PCR. *Lab on a Chip, 10*(21), 2841–2843.

70. Eastburn, D. J., Sciambi, A., & Abate, A. R. (2013). Ultrahigh-throughput mammalian single-cell reverse-transcriptase polymerase chain reaction in microfluidic drops. *Analytical Chemistry, 85*(16), 8016–8021.

71. Tao, Y., et al. (2015). Rapid, targeted and culture-free viral infectivity assay in drop-based microfluidics. *Lab on a Chip, 15*(19), 3934–3940.

72. Mazutis, L., et al. (2013). Single-cell analysis and sorting using droplet-based microfluidics. *Nature Protocols, 8*(5), 870–891.

73. Sarkar, S., et al. (2015). T cell dynamic activation and functional analysis in nanoliter droplet microarray. *Journal of Clinical & Cellular Immunology, 6*(3), 334.

74. Han, Q., et al. (2012). Polyfunctional responses by human T cells result from sequential release of cytokines. *Proceedings of the National Academy of Sciences, 109*(5), 1607–1612.

75. Fan, H. C., Fu, G. K., & Fodor, S. P. (2015). Expression profiling: Combinatorial labeling of single cells for gene expression cytometry. *Science, 347*(6222), 1258367.

76. Yamanaka, Y. J., et al. (2012). Cellular barcodes for efficiently profiling single-cell secretory responses by microengraving. *Analytical Chemistry, 84*(24), 10531–10536.

77. Ogunniyi, A. O., et al. (2009). Screening individual hybridomas by microengraving to discover monoclonal antibodies. *Nature Protocols, 4*(5), 767–782.

78. Kobel, S., et al. (2010). Optimization of microfluidic single cell trapping for long-term on-chip culture. *Lab on a Chip, 10*(7), 857–863.

79. Jin, D., et al. (2015). A microfluidic device enabling high-efficiency single cell trapping. *Biomicrofluidics, 9*(1), 014101.

80. Kimmerling, R. J., et al. (2016). A microfluidic platform enabling single-cell RNA-seq of multigenerational lineages. *Nature Communications, 7*, 10220.

81. Di Carlo, D., Aghdam, N., & Lee, L. P. (2006). Single-cell enzyme concentrations, kinetics, and inhibition analysis using high-density hydrodynamic cell isolation arrays. *Analytical Chemistry, 78*(14), 4925–4930.

82. Luan, Q., et al. (2020). Microfluidic systems for hydrodynamic trapping of cells and clusters. *Biomicrofluidics, 14*(3), 031502.

83. Kirby, B. J. (2010). *Micro- and nanoscale fluid mechanics: Transport in microfluidic devices.* Cambridge University Press.

84. Tan, W. H., & Takeuchi, S. (2007). A trap-and-release integrated microfluidic system for dynamic microarray applications. *Proceedings of the National Academy of Sciences, 104*(4), 1146–1151.

85. Frimat, J. P., et al. (2011). A microfluidic array with cellular valving for single cell co-culture. *Lab on a Chip, 11*(2), 231–237.

86. Abedini-Nassab, R. (2019). Magnetomicrofluidic platforms for organizing arrays of single-particles and particle-pairs. *Journal of Microelectromechanical Systems, 28*(4), 732–738.

87. Skelley, A. M., et al. (2009). Microfluidic control of cell pairing and fusion. *Nature Methods, 6*(2), 147–152.

88. Liu, Y., et al. (1995). Evidence for localized cell heating induced by infrared optical tweezers. *Biophysical Journal, 68*(5), 2137–2144.

89. Bettinger, C., Borenstein, J. T., & Tao, S. L. (2012). *Microfluidic cell culture systems.* Elsevier Science.

90. Fabbri, F., et al. (2013). Detection and recovery of circulating colon cancer cells using a dielectrophoresis-based device: KRAS mutation status in pure CTCs. *Cancer Letters, 335*(1), 225–231.

91. Fritzsch, F. S., et al. (2013). Picoliter nDEP traps enable time-resolved contactless single bacterial cell analysis in controlled microenvironments. *Lab on a Chip, 13*(3), 397–408.

92. Mittal, N., Rosenthal, A., & Voldman, J. (2007). nDEP microwells for single-cell patterning in physiological media. *Lab on a Chip, 7*(9), 1146–1153.

93. Peeters, D. J., et al. (2013). Semiautomated isolation and molecular characterisation of single or highly purified tumour cells from cellsearch enriched blood samples using dielectrophoretic cell sorting. *British Journal of Cancer, 108*(6), 1358–1367.

94. Neuman, K. C., & Block, S. M. (2004). Optical trapping. *Review of Scientific Instruments, 75*(9), 2787–2809.

95. Jing, P., et al. (2016). Photonic crystal optical tweezers with high efficiency for live biological samples and viability characterization. *Scientific Reports, 6*, 19924.

96. Chiou, P. Y., Ohta, A. T., & Wu, M. C. (2005). Massively parallel manipulation of single cells and microparticles using optical images. *Nature, 436*(7049), 370–372.

97. Collins, D. J., et al. (2015). Two-dimensional single-cell patterning with one cell per well driven by surface acoustic waves. *Nature Communications, 6*, 8686.

98. Marx, V. (2015). Biophysics: Using sound to move cells. *Nature Methods, 12*(1), 41–44.

99. Ding, X., et al. (2012). On-chip manipulation of single microparticles, cells, and organisms using surface acoustic waves. *Proceedings of National Academy of Sciences, 109*(28), 11105–11109.

100. Guo, F., et al. (2016). Three-dimensional manipulation of single cells using surface acoustic waves. *Proceedings of the National Academy of Sciences, 113*(6), 1522–1527.

101. Lee, H., et al. (2007). Integrated cell manipulation system—CMOS/microfluidic hybrid. *Lab on a Chip, 7*(3), 331–337.

102. Henighan, T., et al. (2010). Manipulation of magnetically labeled and unlabeled cells with mobile magnetic traps. *Biophysical Journal, 98*(3), 412–417.

103. Donolato, M., et al. (2011). Magnetic domain wall conduits for single cell applications. *Lab on a Chip, 11*(17), 2976–2983.

104. Liu, W., et al. (2009). A novel permalloy based magnetic single cell micro array. *Lab on a Chip, 9*(16), 2381–2390.

105. Pohl, H. A. (1951). The motion and precipitation of suspensoids in divergent electric fields. *Journal of Applied Physics, 22*(7), 869–871.

106. Arnold, M. S., et al. (2006). Sorting carbon nanotubes by electronic structure using density differentiation. *Nature Nanotechnology, 1*(1), 60–65.

107. Krupke, R., et al. (2003). Separation of metallic from semiconducting single-walled carbon nanotubes. *Science, 301*(5631), 344–347.
108. Kullock, R., et al. (2020). Electrically-driven Yagi-Uda antennas for light. *Nature Communications, 11*(1), 115.
109. Lorenz, M., et al. (2020). Aerosol classification by dielectrophoresis: A theoretical study on spherical particles. *Scientific Reports, 10*(1), 10617.
110. Becker, F. F., et al. (1995). Separation of human breast cancer cells from blood by differential dielectric affinity. *Proceedings of the National Academy of Sciences, 92*(3), 860–864.
111. Voldman, J. (2006). Electrical forces for microscale cell manipulation. *Annual Review of Biomedical Engineering, 8*, 425–454.
112. Fiedler, S., et al. (1998). Dielectrophoretic sorting of particles and cells in a microsystem. *Analytical Chemistry, 70*(9), 1909–1915.
113. Pohl, H. A., & Hawk, I. (1966). Separation of living and dead cells by dielectrophoresis. *Science, 152*(3722), 647–649.
114. Sang, S., et al. (2016). Portable microsystem integrates multifunctional dielectrophoresis manipulations and a surface stress biosensor to detect red blood cells for hemolytic anemia. *Scientific Reports, 6*, 33626.
115. Yu, E. S., et al. (2020). Precise capture and dynamic relocation of nanoparticulate biomolecules through dielectrophoretic enhancement by vertical nanogap architectures. *Nature Communications, 11*(1), 2804.
116. Green, N. G., Morgan, H., & Milner, J. J. (1997). Manipulation and trapping of sub-micron bioparticles using dielectrophoresis. *Journal of Biochemical and Biophysical Methods, 35*(2), 89–102.
117. Hamada, R., et al. (2013). A rapid bacteria detection technique utilizing impedance measurement combined with positive and negative dielectrophoresis. *Sensors and Actuators B: Chemical, 181*, 439–445.
118. Madiyar, F. R., et al. (2013). Manipulation of bacteriophages with dielectrophoresis on carbon nanofiber nanoelectrode arrays. *Electrophoresis, 34*(7), 1123–1130.
119. Lin, Y., Shiomi, J., & Amberg, G. (2009). Numerical calculation of the dielectrophoretic force on a slender body. *Electrophoresis, 30*(5), 831–838.
120. Techaumnat, B., Eua-arporn, B., & Takuma, T. (2004). Calculation of electric field and dielectrophoretic force on spherical particles in chain. *Journal of Applied Physics, 95*(3), 1586–1593.
121. Abedini-Nassab, R., et al. (2022). Quantifying the dielectrophoretic force on colloidal particles in microfluidic devices. *Microfluidics and Nanofluidics, 26*(5), 38.
122. Crews, N., et al. (2007). An analysis of interdigitated electrode geometry for dielectrophoretic particle transport in micro-fluidics. *Sensors and Actuators B: Chemical, 125*(2), 672–679.
123. Javanmard, M., et al. (2012). Use of negative dielectrophoresis for selective elution of protein-bound particles. *Analytical Chemistry, 84*(3), 1432–1438.
124. Yan, S., et al. (2014). On-chip high-throughput manipulation of particles in a dielectrophoresis-active hydrophoretic focuser. *Scientific Reports, 4*, 5060.
125. Albrecht, D. R., Sah, R. L., & Bhatia, S. N. (2004). Geometric and material determinants of patterning efficiency by dielectrophoresis. *Biophysical Journal, 87*(4), 2131–2147.
126. Morgan, H., & Green, N. G. (1997). Dielectrophoretic manipulation of rod-shaped viral particles. *Journal of Electrostatics, 42*(3), 279–293.
127. Sedgwick, H., et al. (2008). Lab-on-a-chip technologies for proteomic analysis from isolated cells. *Journal of the Royal Society Interface, 5*(Suppl 2), S123–130.
128. Rosenthal, A., & Voldman, J. (2005). Dielectrophoretic traps for single-particle patterning. *Biophysical Journal, 88*(3), 2193–2205.
129. Fuhr, G., et al. (1992). Levitation, holding, and rotation of cells within traps made by high-frequency fields. *Biochimica et Biophysica Acta, 1108*(2), 215–223.
130. Voldman, J., et al. (2003). Design and analysis of extruded quadrupolar dielectrophoretic traps. *Journal of Electrostatics, 57*(1), 69–90.

131. Schnelle, T., et al. (1993). Three-dimensional electric field traps for manipulation of cells—Calculation and experimental verification. *Biochimica et Biophysica Acta, 1157*(2), 127–140.
132. Schnelle, T., Müller, T., & Fuhr, G. (2000). Trapping in AC octode field cages. *Journal of Electrostatics, 50*(1), 17–29.
133. Gerard, H. M., Ronald, P., & Juliette, R. (1997). The dielectrophoretic levitation of latex beads, with reference to field-flow fractionation. *Journal of Physics D: Applied Physics, 30*(17), 2470.
134. Cheng, I. F., et al. (2009). A continuous high-throughput bioparticle sorter based on 3D traveling-wave dielectrophoresis. *Lab on a Chip, 9*(22), 3193–3201.
135. Cui, H. H., et al. (2009). Separation of particles by pulsed dielectrophoresis. *Lab on a Chip, 9*(16), 2306–2312.
136. Wang, L., et al. (2009). Dual frequency dielectrophoresis with interdigitated sidewall electrodes for microfluidic flow-through separation of beads and cells. *Electrophoresis, 30*(5), 782–791.
137. Urdaneta, M., & Smela, E. (2007). Multiple frequency dielectrophoresis. *Electrophoresis, 28*(18), 3145–3155.
138. Ngo, T.-T., et al. (2014). A planar interdigital sensor for bio-impedance measurement: Theoretical analysis, optimization and simulation. *Journal Nano- and Electronic Physics, 6*(1), 01011 (7pp).
139. Rahman, M. R. U., et al. (2021). Effect of geometry on dielectrophoretic trap stiffness in microparticle trapping. *Biomedical Microdevices, 23*(3), 33.
140. Voldman, J., et al. (2001). Holding forces of single-particle dielectrophoretic traps. *Biophysical Journal, 80*(1), 531–541.
141. Nejad, H. R., et al. (2013). Characterization of the geometry of negative dielectrophoresis traps for particle immobilization in digital microfluidic platforms. *Lab on a Chip, 13*(9), 1823–1830.
142. Saucedo-Espinosa, M. A., & Lapizco-Encinas, B. H. (2015). Experimental and theoretical study of dielectrophoretic particle trapping in arrays of insulating structures: Effect of particle size and shape. *Electrophoresis, 36*(9–10), 1086–1097.
143. Kwak, T. J., et al. (2021). Size-selective particle trapping in dielectrophoretic corral traps. *The Journal of Physical Chemistry C, 125*(11), 6278–6286.
144. Ramos, A., et al. (1999). AC electric-field-induced fluid flow in microelectrodes. *Journal of Colloid and Interface Science, 217*(2), 420–422.
145. Castellanos, A., et al. (2003). Electrohydrodynamics and dielectrophoresis in microsystems: Scaling laws. *Journal of Physics D: Applied Physics, 36*(20), 2584.
146. Gascoyne, P. R., & Vykoukal, J. (2002). Particle separation by dielectrophoresis. *Electrophoresis, 23*(13), 1973–1983.
147. Pethig, R. (2010). Review article-dielectrophoresis: Status of the theory, technology, and applications. *Biomicrofluidics, 4*(2), 022811.
148. Gagnon, Z. R. (2011). Cellular dielectrophoresis: Applications to the characterization, manipulation, separation and patterning of cells. *Electrophoresis, 32*(18), 2466–2487.
149. Ramos, A., et al. (1998). Ac electrokinetics: A review of forces in microelectrode structures. *Journal of Physics D: Applied Physics, 31*(18), 2338.
150. Vallet, M., Berge, B., & Vovelle, L. (1996). Electrowetting of water and aqueous solutions on poly(ethylene terephthalate) insulating films. *Polymer, 37*(12), 2465–2470.
151. Choi, K., et al. (2012). Digital microfluidics. *Annual Review of Analytical Chemistry, 5*(1), 413–440.
152. Huang, S., et al. (2020). Digital microfluidics for the detection of selected inorganic ions in aerosols. *Sensors, 20*(5), 1281.
153. Grenier, N., Merville, P., & Combe, C. (2016). Radiologic imaging of the renal parenchyma structure and function. *Nature Reviews Nephrology, 12*(6), 348–359.
154. Li, L., et al. (2013). Superparamagnetic iron oxide nanoparticles as MRI contrast agents for non-invasive stem cell labeling and tracking. *Theranostics, 3*(8), 595–615.

155. Qiao, R., Yang, C., & Gao, M. (2009). Superparamagnetic iron oxide nanoparticles: From preparations to in vivo MRI applications. *Journal of Materials Chemistry, 19*(35), 6274–6293.

156. Ahrens, E. T., & Bulte, J. W. (2013). Tracking immune cells in vivo using magnetic resonance imaging. *Nature Reviews Immunology, 13*(10), 755–763.

157. Neuwelt, A., et al. (2015). Iron-based superparamagnetic nanoparticle contrast agents for MRI of infection and inflammation. *AJR American Journal of Roentgenology, 204*(3), W302–313.

158. Johnson, L. M., et al. (2014). Multiparametric MRI in prostate cancer management. *Nature Reviews Clinical Oncology, 11*(6), 346–353.

159. Wang, G., et al. (2016). Au nanocage functionalized with ultra-small Fe_3O_4 nanoparticles for targeting T1–T2dual MRI and CT imaging of tumor. *Scientific Reports, 6*, 28258.

160. Li, H., et al. (2016). Quantitative MRI radiomics in the prediction of molecular classifications of breast cancer subtypes in the TCGA/TCIA data set. *NPJ Breast Cancer, 2*, 16012.

161. Liu, H., et al. (2016). A semi-mechanism approach based on MRI and proteomics for prediction of conversion from mild cognitive impairment to Alzheimer's disease. *Scientific Reports, 6*, 26712.

162. Gleich, B., & Weizenecker, J. (2005). Tomographic imaging using the nonlinear response of magnetic particles. *Nature, 435*(7046), 1214–1217.

163. Pankhurst, Q. A., et al. (2009). Progress in applications of magnetic nanoparticles in biomedicine. *Journal of Physics D: Applied Physics, 42*(22), 224001.

164. Weizenecker, J., et al. (2009). Three-dimensional real-time in vivo magnetic particle imaging. *Physics in Medicine & Biology, 54*(5), L1–L10.

165. Them, K., et al. (2016). Increasing the sensitivity for stem cell monitoring in system-function based magnetic particle imaging. *Physics in Medicine & Biology, 61*(9), 3279–3290.

166. Borgert, J., et al. (2012). Fundamentals and applications of magnetic particle imaging. *Journal of Cardiovascular Computed Tomography, 6*(3), 149–153.

167. Freeman, M. W., Arrott, A., & Watson, J. H. L. (1960). Magnetism in medicine. *Journal of Applied Physics, 31*(5), S404–S405.

168. Huang, J., et al. (2016). Magnetic nanoparticle facilitated drug delivery for cancer therapy with targeted and image-guided approaches. *Advanced Functional Materials, 26*(22), 3818–3836.

169. Chen, M. L., et al. (2012). Quantum dots conjugated with Fe_3O_4-filled carbon nanotubes for cancer-targeted imaging and magnetically guided drug delivery. *Langmuir, 28*(47), 16469–16476.

170. Eguchi, H., et al. (2015). A magnetic anti-cancer compound for magnet-guided delivery and magnetic resonance imaging. *Scientific Reports, 5*, 9194.

171. Lubbe, A. S., et al. (1996). Clinical experiences with magnetic drug targeting: A phase I study with 4'-epidoxorubicin in 14 patients with advanced solid tumors. *Cancer Research, 56*(20), 4686–4693.

172. Jordan, A., et al. (1999). Magnetic fluid hyperthermia (MFH): Cancer treatment with AC magnetic field induced excitation of biocompatible superparamagnetic nanoparticles. *Journal of Magnetism and Magnetic Materials, 201*(1–3), 413–419.

173. Kim, M. H., et al. (2013). Magnetic nanoparticle targeted hyperthermia of cutaneous Staphylococcus aureus infection. *Annals of Biomedical Engineering, 41*(3), 598–609.

174. Williams, J. P., et al. (2013). Application of magnetic field hyperthermia and superparamagnetic iron oxide nanoparticles to HIV-1-specific T-cell cytotoxicity. *International Journal of Nanomedicine, 8*, 2543–2454.

175. Mamiya, H., & Jeyadevan, B. (2011). Hyperthermic effects of dissipative structures of magnetic nanoparticles in large alternating magnetic fields. *Scientific Reports, 1*, 157.

176. Branquinho, L. C., et al. (2013). Effect of magnetic dipolar interactions on nanoparticle heating efficiency: Implications for cancer hyperthermia. *Scientific Reports, 3*, 2887.

177. Che Rose, L., et al. (2016). A SPION-eicosane protective coating for water soluble capsules: Evidence for on-demand drug release triggered by magnetic hyperthermia. *Scientific Reports, 6*, 20271.

178. Regmi, R., et al. (2010). Hyperthermia controlled rapid drug release from thermosensitive magnetic microgels. *Journal of Materials Chemistry, 20*(29), 6158–6163.

179. Lee, H., et al. (2010). High-speed RNA microextraction technology using magnetic oligo-dT beads and lateral magnetophoresis. *Lab on a Chip, 10*(20), 2764–2770.
180. He, X., et al. (2007). Plasmid DNA isolation using amino-silica coated magnetic nanoparticles (ASMNPs). *Talanta, 73*(4), 764–769.
181. Rittich, B., et al. (2006). Isolation of microbial DNA by newly designed magnetic particles. *Colloids and Surfaces B: Biointerfaces, 52*(2), 143–148.
182. Min, J. H., et al. (2014). Isolation of DNA using magnetic nanoparticles coated with dimercaptosuccinic acid. *Analytical Biochemistry, 447*, 114–118.
183. He, J., et al. (2014). Magnetic separation techniques in sample preparation for biological analysis: A review. *Journal of Pharmaceutical and Biomedical Analysis, 101*, 84–101.
184. Schafer, F., et al. (2002). Automated high-throughput purification of 6xHis-tagged proteins. *Journal of Biomolecular Technology, 13*(3), 131–142.
185. Safarik, I., & Safarikova, M. (2004). Magnetic techniques for the isolation and purification of proteins and peptides. *Biomagnetic Research and Technology, 2*(1), 7.
186. Cato, M. H., Yau, I. W., & Rickert, R. C. (2011). Magnetic-based purification of untouched mouse germinal center B cells for ex vivo manipulation and biochemical analysis. *Nature Protocols, 6*(7), 953–960.
187. Chen, P., et al. (2015). Microscale magnetic field modulation for enhanced capture and distribution of rare circulating tumor cells. *Science and Reports, 5*, 8745.
188. Zborowski, M., et al. (1995). Analytical magnetapheresis of ferritin-labeled lymphocytes. *Analytical Chemistry, 67*(20), 3702–3712.
189. Chen, P., et al. (2014). Multiscale immunomagnetic enrichment of circulating tumor cells: From tubes to microchips. *Lab on a Chip, 14*(3), 446–458.
190. Huang, Y. Y., et al. (2015). Screening and molecular analysis of single circulating tumor cells using micromagnet array. *Scientific Reports, 5*, 16047.
191. Li, Y., et al. (2014). Asynchronous Magnetic Bead Rotation (AMBR) microviscometer for label-free DNA analysis. *Biosensors, 4*(1), 76–89.
192. McNaughton, B. H., et al. (2007). Single bacterial cell detection with nonlinear rotational frequency shifts of driven magnetic microspheres. *Applied Physics Letters, 91*(22), 224105.
193. Sinn, I., et al. (2011). Asynchronous magnetic bead rotation (AMBR) biosensor in microfluidic droplets for rapid bacterial growth and susceptibility measurements. *Lab on a Chip, 11*(15), 2604–2611.
194. Sinn, I., et al. (2012). Asynchronous magnetic bead rotation microviscometer for rapid, sensitive, and label-free studies of bacterial growth and drug sensitivity. *Analytical Chemistry, 84*(12), 5250–5256.
195. Kinnunen, P., et al. (2011). Monitoring the growth and drug susceptibility of individual bacteria using asynchronous magnetic bead rotation sensors. *Biosensors and Bioelectronics, 26*(5), 2751–2755.
196. Hecht, A., et al. (2011). Label-acquired magnetorotation for biosensing: An asynchronous rotation assay. *Journal of Magnetism and Magnetic Materials, 323*(3–4), 272–278.
197. Binasch, G., et al. (1989). Enhanced magnetoresistance in layered magnetic structures with antiferromagnetic interlayer exchange. *Physical Review B Condens Matter, 39*(7), 4828–4830.
198. Ramli, R., et al. (2011). GMR biosensors for clinical diagnostics. *Biosensors for health, environment and biosecurity*. IntechOpen.
199. Klein, T., et al. (2019). Development of a multiplexed giant magnetoresistive biosensor array prototype to quantify ovarian cancer biomarkers. *Biosensors and Bioelectronics, 126*, 301–307.
200. Ng, E., et al. (2020). Early multiplexed detection of cirrhosis using giant magnetoresistive biosensors with protein biomarkers. *ACS Sensors, 5*(10), 3049–3057.
201. Reig, C., Cubells-Beltran, M. D., & Munoz, D. R. (2009). Magnetic field sensors based on giant magnetoresistance (GMR) technology: Applications in electrical current sensing. *Sensors, 9*(10), 7919–7942.
202. Li, G., et al. (2006). Spin valve sensors for ultrasensitive detection of superparamagnetic nanoparticles for biological applications. *Sensors and Actuators A: Physical, 126*(1), 98–106.

203. Devkota, J., et al. (2015). A novel approach for detection and quantification of magnetic nanomarkers using a spin valve GMR-integrated microfluidic sensor. *RSC Advances, 5*(63), 51169–51175.
204. Lei, H., et al. (2016). Contactless measurement of magnetic nanoparticles on lateral flow strips using tunneling magnetoresistance (TMR) sensors in differential configuration. *Sensors, 16*(12), 2130.
205. Wu, Y., et al. (2017). Rapid detection of Escherichia coli O157:H7 using tunneling magnetoresistance biosensor. *AIP Advances, 7*(5), 056658.
206. Mu, X.-H., et al. (2019). A new rapid detection method for ricin based on tunneling magnetoresistance biosensor. *Sensors and Actuators B: Chemical, 284*, 638–649.
207. Sharma, P. P., et al. (2017). Integrated platform for detecting pathogenic DNA via magnetic tunneling junction-based biosensors. *Sensors and Actuators B: Chemical, 242*, 280–287.
208. Jaufenthaler, A., et al. (2021). Pulsed optically pumped magnetometers: Addressing dead time and bandwidth for the unshielded magnetorelaxometry of magnetic nanoparticles. *Sensors, 21*(4), 1212.
209. Wiekhorst, F., et al. (2012). Magnetorelaxometry assisting biomedical applications of magnetic nanoparticles. *Pharmaceutical Research, 29*(5), 1189–1202.
210. Huang, C.-C., Zhou, X., & Hall, D. A. (2017). Giant magnetoresistive biosensors for time-domain magnetorelaxometry: A theoretical investigation and progress toward an immunoassay. *Scientific Reports, 7*(1), 45493.
211. Flynn, E. R. (2019). Magnetic relaxometry: A comparison to magnetoencephalography. In S. Supek & C. J. Aine (Eds.), *Magnetoencephalography: From signals to dynamic cortical networks* (pp. 1343–1355). Springer International Publishing.
212. Buchner, M., et al. (2018). Tutorial: Basic principles, limits of detection, and pitfalls of highly sensitive SQUID magnetometry for nanomagnetism and spintronics. *Journal of Applied Physics, 124*(16), 161101.
213. Enpuku, K., et al. (1999). Detection of magnetic nanoparticles with superconducting quantum interference device (SQUID) magnetometer and application to immunoassays. *Japanese Journal of Applied Physics, 38*(Part 2, No. 10A), L1102–L1105.
214. Škrátek, M., et al. (2020). Sensitive SQUID bio-magnetometry for determination and differentiation of biogenic iron and iron oxide nanoparticles in the biological samples. *Nanomaterials, 10*(10), 1993.
215. Min, C., et al. (2017). Integrated microHall magnetometer to measure the magnetic properties of nanoparticles. *Lab on a Chip, 17*(23), 4000–4007.
216. Jogschies, L., et al. (2015). Recent developments of magnetoresistive sensors for industrial applications. *Sensors, 15*(11), 28665–28689.
217. Hien, L. T., et al. (2016). DNA-magnetic bead detection using disposable cards and the anisotropic magnetoresistive sensor. *Advances in Natural Sciences: Nanoscience and Nanotechnology, 7*(4), 045006.

Chapter 2
Circuit Theory

Electrical and electronic circuits are now well developed and used almost everywhere (e.g., in our computer systems, mobile phones, control systems, and so on). Using the idea of circuits, precise control of electrical currents is achieved, and thus, storing a huge amount of data in computer memory chips is now possible. These datasets are then processed and transported using microprocessors, which again work based on the circuit concept and capability of controlling numerous electrical currents and voltages in parallel.

Now, in single-cell biology, achieving a similar control on single cells, as opposed to electrons in electrical circuits, allows us to deal with a large number of single cells. To achieve the highly scalable, automated, and programmable functionality of the integrated circuits, which is not fully seen in typical single-cell analysis tools (introduced in the first chapter), a circuit-based approach can play a key role. The circuits can be modified for transporting magnetic particles and magnetized single cells within microfluidic devices, which act analogous to the ones in computer chips. Thus, before discussing these circuits, called magnetophoretic circuits, an introduction to the circuit theory terms and concepts would be helpful.

2.1 Electrical and Electronic Circuits

Today, electrical and electronic circuits have fundamentally enhanced modern life with systems for delivering energy (power electronics), delivering data (telecommunication), solving complex problems (computation), monitoring processes, and controlling the systems. But three key achievements are the reasons behind achieving these diverse functionalities: (i) passive and active circuit elements (*e.g.*, resistors, capacitors, inductors, diodes, and transistors) are designed, (ii) the electron transport phenomena through these elements are understood, and (iii) the circuit elements are integrated to design complex integrated circuits with key functionalities.

© The Author(s), under exclusive license to Springer Nature Singapore Pte Ltd. 2023
R. Abedini-Nassab, *Magnetomicrofluidic Circuits for Single-Bioparticle Transport*,
https://doi.org/10.1007/978-981-99-1702-0_2

Resistors obey Ohm's law, which is a consequence of the charge conservation laws and the constitutive relationships of materials. Based on Ohm's law, the current through a resistor is directly proportional to the applied voltage to its terminals (*i.e.*, $I = \sigma V$, where I, σ, and V are the current through the resistor, its conductivity, and the voltage across it, respectively). Kirchhoff's current and voltage laws are also used, in addition to Ohm's law, for analyzing electrical circuits. Based on these rules, the algebraic sum of the currents entering a node and the algebraic sum of the electric potentials along a closed path are zero.

Capacitors store electrical charges. This accumulated electrical charge is proportional to the applied electric potential (*i.e.*, $Q = CV$, where Q and C are the electric charge and capacitance, respectively). Also, inductors store energy in them. A change in the current passing through the inductor results in a time-varying magnetic field which induces an electromotive force. Based on Lenz's law, the polarity of the induced voltage opposes the current change that produced it. The magnetic flux in inductors is proportional to the electrical current (*i.e.*, $\Phi = LI$, where Φ and L are the magnetic flux and inductance, respectively). Diodes are electronic circuit elements that conduct electrical currents in one direction only. That means when they are biased in forward mode, current moves through them. But, in a reversed bias (*i.e.*, applying a voltage with inversed polarity compared to the one in the forward bias), the electrical current passing through the device is theoretically zero.

The resistors, capacitors, diodes, and inductors are passive circuit elements that are used to distribute the required signals to the systems. But, in order to switch the electrical currents (*i.e.*, the electron transports) at a specific spot on the circuit, transistors are used. Basically, transistors switch the electrical signal based on an input control signal. Transistors are the key circuit elements in electronic circuits, including computer chips.

Although computers nowadays are well developed, they have a long history back to 1936. At that time, Alan Turing theoretically introduced a problem-solving machine that worked on simple instructions encoded on a tape. That pioneering work then became the foundation of the computer world and the computational sciences [1].

Computer systems are extremely large collections of electrical switches, which control and manipulate various data based on their input signals. In computer memory chips, these switches are arranged in developed memory architectures, forming banks, rows, and columns, to dramatically reduce the required control wires. The switches in early computers were vacuum tubes (triodes), invented by Lee De Forest, in 1906. In vacuum tubes, a control grid separates a cathode and a plate. Cathode gets heated and emits electrons which are then collected by the plate. But the electron flow rate is regulated by the control grid.

Later, by introducing the solid-state circuits, the transistors replaced the vacuum tubes. Transistors are considered more reliable switches compared to vacuum tubes. Transistors are mainly categorized into two groups of Bipolar Junction Transistors (BJT) and Metal Oxide Semiconductor Field Effect Transistors (MOSFET). In BJTs, a control signal applied to the base terminal defines the emitter-collector electrical current. Similarly, in MOSFETS, the gate control signal regulates the drain-source electrical current.

After the invention of solid-state transistors, Robert Noyce at Fairchild Camera and Jack Kilby at Texas Instruments developed the idea of integrating all the circuit elements on a single substrate, resulting in what is called integrated circuits (IC). At that time, the ICs were only composed of a few transistors; however, nowadays computer processors with billions of transistors are commercially available [2]. The integration of numerous transistors on a small chip has become possible by lowering the transistor size, which also results in less power consumption and higher operational speeds. By shrinking the transistor sizes and entering the nanometer range, the transistors are supposed to manipulate single electrons (maybe by single-electron transistors (SET) [3, 4]). Nanowires and quantum dots may play a key role. At that scale, the governing equations are defined based on quantum mechanics (*e.g.*, the Schrodinger equation). Currently, electronic integrated circuits are mostly fabricated on silicon chips, using semiconductor processing, microfabrication, and nanofabrication techniques.

2.2 Optical Circuits

The electronic circuits are well developed. But let's think about the circuit concept in a more general sense. In electrical circuits, the mobile components are the electrons. Now, consider circuits in which the mobile components are objects other than the electrons. For example, photons travel faster than electrons and have lower heat dissipation. Hence, in optical circuits, data transportation can be faster, and less heat may be created. As a result, these circuits can operate at higher speeds, delivering increased computing power. Researchers have designed an optical transistor that based on an input control signal can control the route of a photon from the input to the output [5]. This device operates based on a mirror pair (*i.e.*, microresonator), through which light passes or does not pass when the switch is on or off, respectively. Similar works are also done on optical circuits [6, 7], and we may one day have an all-optical computer system.

2.3 Magnetophoretic Circuits

Drawing inspiration from electronic circuits, magnetophoretic circuits are developed to precisely manipulate magnetic particles and single cells in a microfluidic environment [8, 9]. The building blocks of these circuits, similar to electronic circuits, are the conductors, diodes, capacitors, and transistors.

The magnetophoretic circuit elements are constructed from lithographically patterned, overlaid magnetic and conductive thin films. These circuits transport magnetized particles by the force created by the externally applied rotating magnetic field. Magnetized particles in the proximity of the magnetic thin films exposed to an in-plane rotating magnetic field move along the magnetic tracks in a manner similar to Ohm's law of electrical circuits. Also, analogous to the diodes in electronic circuits, magnetophoretic diodes are designed to transport magnetized particles and cells in one direction only. Magnetic capacitors store the magnetized particles, similar to what capacitors do for the electrons.

The magnetophoretic transistors, composed of overlaid magnetic thin films and microwires, are designed to switch the particle trajectories to the desired one. This important need is achieved by applying electrical control signals to their gate. Integrating these circuit elements results in a complete magnetophoretic circuit with unique specifications. (i) By using resistors, they transport a large number of single particles and cells in parallel, all synched with the general external rotating magnetic field. This achievement is fundamentally important since it eliminates the need of applying individual signals specific to individual particles, which is a drawback of some other particle manipulation methods. (ii) The magnetophoretic transistors provide the opportunity to precisely control the trajectory of single particles and cells at specific spots. Thus, by combining these circuit elements, precise transport of individual cells in an automated fashion, without requiring complex electrode and wiring systems, is achieved. After providing the required theory background, these circuits will be discussed in detail in the next chapters.

In conclusion, since magnetomicrofluidic chips operate based on the concepts of the circuit theory, this theory is discussed in this chapter. Electrical circuits are the most widely used circuits that have resulted in fundamental advances in the world. They are used in computer systems, mobile phones, smart devices, etc. But in addition to the electrical circuits, with electrons as the mobile components, optical circuits based on photons are introduced. Finally, in this section, the idea of developing other circuits with different mobile components is introduced. It was explained that by drawing inspiration from electrical circuits, magnetophoretic circuits are developed. These circuits offer controlled transport of magnetic particles and single cells in a microfluidic environment.

References

1. Turing, A. M. (1937). On computable numbers, with an application to the Entscheidungsproblem. *Proceedings of the London Mathematical Society, s2–42*(1), 230–265.
2. Alcorn, P. (2016). *Intel Xeon E5–2600 v4 Broadwell-EP review*. http://www.tomshardware.com/reviews/intel-xeon-e5-2600-v4-broadwell-ep,4514-2.html
3. Cheng, G., et al. (2011). Sketched oxide single-electron transistor. *Nature Nanotechnology, 6*(6), 343–347.
4. Kano, S., et al. (2015). Chemically assembled double-dot single-electron transistor analyzed by the orthodox model considering offset charge. *Journal of Applied Physics, 118*(13), 134304.
5. Shomroni, I., et al. (2014). Quantum optics. All-optical routing of single photons by a one-atom switch controlled by a single photon. *Science, 345*(6199), 903–906.
6. Malishava, M., & Khomeriki, R. (2015). All-phononic digital transistor on the basis of gap-soliton dynamics in an anharmonic oscillator ladder. *Physical Review Letters, 115*(10), 104301.
7. Le Kien, F., & Rauschenbeutel, A. (2016). Nanofiber-based all-optical switches. *Physical Review A, 93*(1), 013849.

8. Lim, B., et al. (2014). Magnetophoretic circuits for digital control of single particles and cells. *Nature Communications, 5*, 3846.

9. Abedini-Nassab, R., et al. (2016). Magnetophoretic conductors and diodes in a 3D magnetic field. *Advanced Functional Materials, 26*(22), 4026–4034.

Chapter 3
Theory and Simulation Methods

In this chapter, I will provide the required theoretical background for studying magnetophoretic circuits. First, magnetic materials will be briefly discussed, and then, the theory for calculating the magnetic forces and predicting the magnetic particle trajectories will be explained.

3.1 Magnetic Materials

All matter, when exposed to a magnetic field, shows magnetic properties, but the magnetization level in most of them is too low to be noticed. Magnetic materials exhibit a much higher magnetization level compared to others. Magnetic materials at the bulk level are usually grouped into three categories: diamagnetic, paramagnetic, and ferromagnetic materials [1]. Additionally, some nanoscale magnetic materials form the superparamagnetic material group [2].

All electrons in diamagnetic materials are paired, leaving no net magnetic moment. In this type of material, electrons exposed to an external magnetic field gain an additional angular momentum, which causes the magnetization of those materials [3]. The induced magnetic moment in these materials opposes the external field. Hence, their susceptibility is negative, resulting in a negative slope in their magnetization plot as a function of the external field (see Fig. 3.1a). The plots in Fig. 3.1 demonstrate the magnetization (M) as a function of the magnetic field intensity (H). Usually, a weak magnetic property is observed in all matter (*e.g.*, quartz and water). But, in perfect diamagnetic materials, called superconductors, this is a strong magnetic property and they completely expel the applied magnetic field [4].

In paramagnetic materials, unpaired electrons in their partially filled electronic orbitals pose a net magnetic moment, which, when exposed to an externally applied magnetic field, is partially aligned in that direction [5]. This behavior leads to a positive susceptibility and a positive slope in their magnetization plot as a function

R. Abedini-Nassab, *Magnetomicrofluidic Circuits for Single-Bioparticle Transport*,
https://doi.org/10.1007/978-981-99-1702-0_3

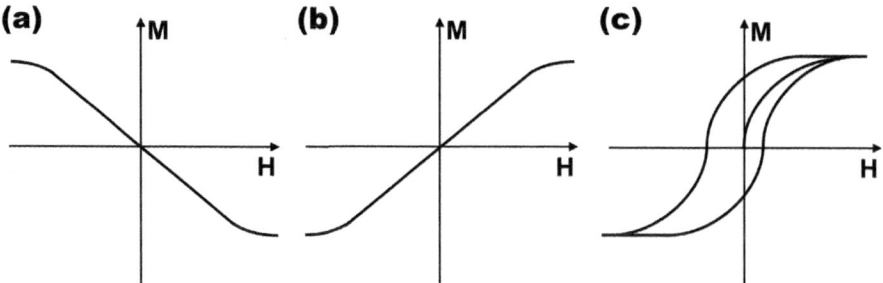

Fig. 3.1 Magnetization plots for (**a**) diamagnetic materials, (**b**) paramagnetic materials, and (**c**) ferromagnetic materials are shown

of the magnetic field (see Fig. 3.1b). Based on Curie Law, the magnetic susceptibility of paramagnetic materials inversely changes with the temperature.

In ferromagnetic materials (e.g., nickel, cobalt, and iron), permanent atomic magnetic dipoles exist even without the application of the external magnetic field. Ferromagnetic materials, when exposed to a magnetic field, respond to it, store magnetization when that field is removed, and exhibit a hysteresis behavior (see Fig. 3.1c). At temperatures above the Curie temperature, the magnetic dipole coupling in them cannot dominate the thermal energy and lose their hysteresis properties. Thus, at high enough temperatures, they behave similarly to a paramagnetic material.

Superparamagnetic materials display ferromagnetic behavior in the bulk state but behave differently at the nanoscale (e.g., metal or metal-oxide nanoparticles). At this size, the thermal energy fluctuation they receive from their surrounding area is strong enough to randomly change their magnetization direction. Thus, they do not need high temperatures (e.g., higher than the Curie point) to lose their hysteresis properties. Superparamagnetic materials behave similarly to paramagnetic materials with extremely higher magnetization levels [6, 7]. Also, in an external magnetic field gradient, they feel a strong magnetic force, and when the field is removed, they show almost no magnetic properties. Hence, in the absence of magnetic fields, they don't aggregate. These two interesting specifications make these materials good candidates for performing biological applications, in which, in the presence and absence of magnetic fields, strong forces, and no attraction forces, respectively, are desired. For example, in magnetophoretic circuits, the moving particles are chosen to be superparamagnetic materials.

3.2 Analytical Modeling

To model the movement of the magnetic particles and magnetized biological cells in the magnetophoretic circuits, the effect of various magnetic fields on them needs to be considered. Here, I will first discuss how magnetic particles respond in an external magnetic field. Then, I will explain the various available magnetic field sources in a magnetophoretic circuit and their effect on the magnetic particles.

3.2.1 Force on Magnetic Particles

Magnetization of superparamagnetic materials at magnetic fields weaker than their saturation point is expressed by Eq. (3.1):

$$\vec{m} = \left(\chi_p - \chi_f\right) V_p \vec{H}, \tag{3.1}$$

where χ_p, χ_f, V_p, and H stand for the magnetic susceptibility of the particle, the magnetic susceptibility of the surrounding fluid (media), the particle volume, and the magnetic field intensity, respectively. The magnetization of these materials in magnetic fields stronger than a critical point becomes saturated. This is not a typical case in magnetic manipulation in magnetophoretic circuits though.

Based on electromagnetic theory, a sphere with linear magnetic properties exposed to weakly inhomogeneous (on the length scale of the particles) magnetic fields behaves similarly to a point dipole. A magnetically labeled cell can also be treated as a magnetic point dipole with magnetic moment expressed as Eq. (3.2):

$$\vec{m}_{\text{cell}} = \left(\chi_p - \chi_f\right) \frac{\psi\left(A_{\text{cell}}\right)\left(d_p\right)}{V_{\text{cell}}} V_{\text{cell}} \vec{H}, \tag{3.2}$$

where ψ, A_{cell}, V_{cell}, and d_p are the cell surface nanoparticle coverage area, cell surface area, cell volume, and the diameter of nanoparticles, respectively. Equation (3.2) can simply be written as Eq. (3.3), which is more similar to Eq. (3.1).

$$\vec{m}_{\text{cell}} = \chi_{\text{cell}} V_{\text{cell}} \vec{H}, \tag{3.3}$$

where $\chi_{\text{cell}} = \left(\chi_p - \chi_f\right)\psi\left(A_{\text{cell}}\right)\left(d_p\right)/V_{\text{cell}}$ is defined as the cell effective susceptibility, which is typically much smaller than the magnetic susceptibility of the commercially available superparamagnetic beads. Now, using the calculated susceptibility of the manipulating particle, the force acting on it, as a magnetic dipole, is calculated as Eq. (3.4).

$$\vec{F}_{\text{mag}} = \mu_0(\vec{m} \cdot \nabla)\vec{H}, \tag{3.4}$$

where μ_0 stands for the magnetic permeability of the vacuum.

As shown in Eq. (3.4), to calculate the magnetic force on the particles, the magnetic field the particle sees is required. In the magnetophoretic circuits, this magnetic field has three components, including the externally applied field, the field produced by the magnetic thin films, and the field produced by embedded microwires. The applied magnetic field is known. But the other two field components are required to be calculated. Hence, in the following sections, I will provide calculation methods for these two magnetic fields.

3.2.2 Magnetic Field Sources

We include three different magnetic field sources: (i) the magnetic coils (H_{ext}), (ii) the magnetization of magnetic thin films (H_{mag}), and (iii) microwires on the chip (H_{wire}). Thus, the total magnetic field at observation point r (the center of the particle) is written as Eq. (3.5).

$$\vec{H}(\vec{r}) = \vec{H}_{\text{ext}} + \vec{H}_{\text{mag}}(\vec{r}) + \vec{H}_{\text{wire}}(\vec{r}). \tag{3.5}$$

In Eq. (3.5), the gradient of the magnetic field intensity is calculated. The first term in Eq. (3.5), H_{ext}, is assumed to be uniform, with zero gradients. Thus, it can be neglected. The second term, H_{out}, is valid when the chip is exposed to an external magnetic field, and the thin films get magnetized. To calculate this term, we can assume fictitious equivalent magnetic charges to be created on the surface of the magnetic thin films. These charges are not real and are defined only to make the modeling easier. The equivalent magnetic charge density on the magnetized magnetic thin films is given by Eq. (3.6).

$$\delta = \left(\vec{H}_{\text{in}} - \vec{H}_{\text{out}} \right) \cdot \hat{n}, \tag{3.6}$$

where H_{in} and H_{out} stand for magnetic fields inside and outside magnetic thin films, respectively, and n is the local outward unit vector normal to the film surface. Integrating all the surface charges is used to calculate the produced magnetic potential, as stated in Eq. (3.7).

$$\varphi_m(\vec{r}) = \frac{1}{4\pi} \oiint \frac{\delta(\vec{r}_s)}{|\vec{r} - \vec{r}_s|} ds, \tag{3.7}$$

where $|r - r_s|$ is the distance between the observation point, r, and the source point, r_s. Now, the magnetic field produced due to the magnetization of the magnetic thin films is calculated by Eq. (3.8).

$$\vec{H}_{\text{mag}}(\vec{r}) = -\nabla \varphi_m(\vec{r}). \tag{3.8}$$

The last term in Eq. (3.5) stands for the magnetic field produced by the current-carrying microwires. If we assume an infinitely thin sheet current, $I_{sh}(r_s)$, this term is estimated using the Biot-Savart law.

$$\vec{H}_{\text{wire}}(\vec{r}) = \frac{1}{4\pi} \oiint \frac{\overrightarrow{I_{sc}}(\vec{r}_s) \times (\vec{r} - \vec{r}_s)}{|\vec{r} - \vec{r}_s|^3} ds. \tag{3.9}$$

Please note that this is not a self-consistent calculation. In other words, the effect of the current-carrying wires on the magnetic thin films and vice versa is not considered.

3.3 Computational Methods

Computing the integrals, such as the one introduced in Eq. (3.7), sometimes becomes a heavy time-consuming task for computers. To overcome this challenge, appropriate assumptions can help. Based on these assumptions, reasonably good approximate results can be achieved. Towards this goal, two different methods are introduced in the next sections.

3.3.1 Semi-Analytical Solution

The magnetic field produced by magnetized thin films in the magnetophoretic circuits can be calculated using a semi-analytical approximation method [6]. Here, the main concern is to prevent introducing the local numerical artifacts, consistent with the shape-induced anisotropy of thin magnetic films, to the local magnetic fields.

Inhomogeneity in magnetization may result in volumetric pole density inside the magnetic thin films. But, the magnetic thin film, which is typically made of permalloy, is assumed to be an infinitely soft magnetic material (*i.e.*, no hysteresis). In other words, when exposed to a magnetic field, it is uniformly magnetized. Hence, we can treat the magnetic films as a surface pole distribution along their exterior, neglecting the interior volumetric pole density. As another reasonable assumption, we can treat the thin films to be infinitely flat. Hence, we can crumble the surface pole distribution of the magnetic thin film into a one-dimensional magnetic line pole distribution on its perimeter. By rotating the external magnetic field, the line pole distribution shifts across the substrate. We will see in the next chapters that one popular magnetophoretic conductor design is composed of serially connected magnetic disks. The magnetic energy simulation results of disks exposed to an external rotating field are depicted in Fig. 3.2.

Magnetic thin films exposed to an in-plane magnetic field are sometimes modeled as line charges on their perimeter. But this assumption neglects the shape-induced demagnetization factors. To overcome this problem, the thin disks connected in series, as is in the magnetophoretic conductor in Fig. 3.2, can be modeled as a linear assembly of overlapped paramagnetic ellipsoids and oblate spheroids. One advantage of this method is that the influence of shape-induced demagnetizing factors on local magnetization is automatically considered. Thus, the problem mentioned above is

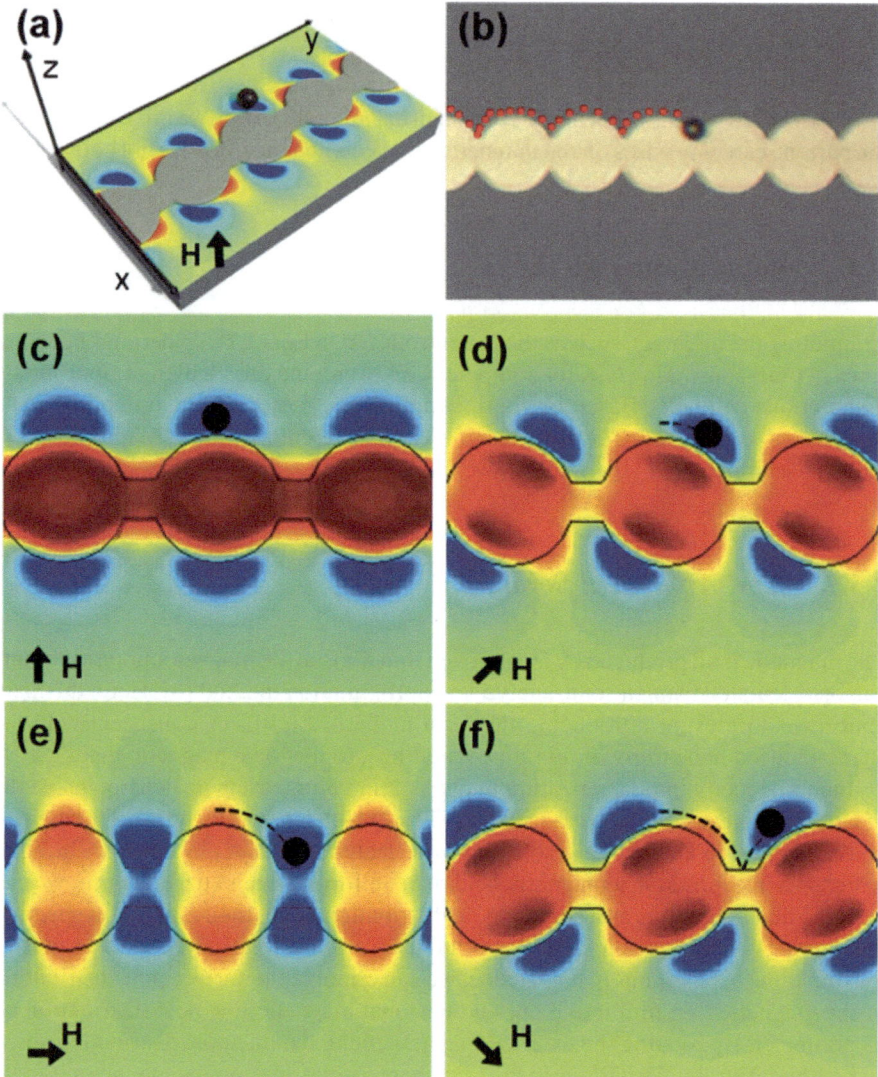

Fig. 3.2 A disk-shaped magnetic thin film in a rotating magnetic field is presented. (**a**) A 3D schematic of a chip with a magnetic particle on it is illustrated. (**b**) A particle trajectory (red dotted line) is overlaid on a microscopy image. (**c–f**) Time series of the energy landscape at the height of particle radius (5 μm) are presented. The black circle, the black dashed line, and the black arrow in each panel stand for the particle, its trajectory, and the magnetic field direction, at each time step, respectively. The disk diameters are 20 μm. © 2019 IEEE. The figure is taken from [8] with permission

solved. But the overlapping regions introduce spurious magnetic pole distributions, which affect local magnetic fields.

A hybrid model is introduced and leads to a semi-self-consistent solution for the local magnetization. It considers the shape-induced demagnetization factors without introducing spurious pole distributions. Towards this goal, the analytical expression for the magnetic potential inside and outside a uniformly magnetized ellipsoid with semi-principal axes of length a, b, and c along the x-, y-, and z-directions, respectively, is used [9–11].

$$\varphi_{out} = \varphi_{ext} + \varphi_{mag} = F_1(\xi)F_2(\eta)F_3(\zeta)\left[C_1 + C_2 \int_\xi^\infty \frac{ds}{R_s(s + a^2)}\right], \tag{3.10}$$

$$\varphi_{in} = C_3 F_1(\xi)F_2(\eta)F_3(\zeta), \tag{3.11}$$

$$R_s = \sqrt{(s + a^2)(s + b^2)(s + c^2)}, \tag{3.12}$$

where φ_{ext} and φ_{mag} are the potentials resulting from the uniform external field aligned in the x-direction and the response field produced by the magnetic ellipsoid, respectively. The coordinate ξ represents the radial term, while η and ζ represent the angular terms, with the characteristic length scale of R_s. In these equations, one-dimensional orthogonal expansions in the ellipsoidal coordinate system are defined as $F1$, $F2$, and $F3$.

$$F_1 = \sqrt{\xi + a^2}, F_2 = \sqrt{\eta + a^2}, F_1 = \sqrt{\zeta + a^2}. \tag{3.13}$$

The coefficients C1, C2, and C3 are calculated by matching the boundary conditions of Eqs. (3.10) and (3.11). The continuity of the magnetic potential and its normal derivative across the boundary are expressed as

$$\phi_{out}|_{\xi=0} = \phi_{in}|_{\xi=0}, \tag{3.14}$$

$$\mu_0\left[\frac{\partial \phi_{out}}{h_\xi \partial \xi}\right]_{\xi=0} = \mu_1\left[\frac{\partial \phi_{in}}{h_\xi \partial \xi}\right]_{\xi=0}, \tag{3.15}$$

where $\xi = 0$ corresponds ellipsoid's surface and $\mu_1 \gg \mu_0$ is the magnetic permeability of the magnetic thin film. The metric coefficient h_ξ is expressed by

$$h_\xi = 0.5\left[\frac{(\xi - \eta)(\xi - \zeta)}{(\xi + a^2)(\xi + b^2)(\xi + c^2)}\right]^{0.5}. \tag{3.16}$$

Now, coefficients C1, C2, and C3 can be inserted into Eqs. (3.10) and (3.11) to get potential outside and inside the ellipsoids.

$$\phi_{\text{out}} = \phi_{\text{out}} \left[\frac{A_a(0, \xi)}{A_a(0, \infty)} + \frac{A_b(0, \xi)}{A_b(0, \infty)} + \frac{A_c(0, \xi)}{A_c(0, \infty)} \right], \tag{3.17}$$

$$\phi_{\text{in}} = - \left[\frac{\vec{H}_{\text{ext}} \cdot \hat{x}}{A_a(0, \infty)} + \frac{\vec{H}_{\text{ext}} \cdot \hat{y}}{A_b(0, \infty)} + \frac{\vec{H}_{\text{ext}} \cdot \hat{z}}{A_c(0, \infty)} \right], \tag{3.18}$$

where A_a, A_b, and A_c are expressed as

$$A_i(p, q) = 1 + \frac{abc}{2} \left(\frac{\mu_1 - \mu_0}{\mu_0} \right) \int_p^q \frac{ds}{(s + i^2) R_s}, \tag{3.19}$$

where p and q take values of 0, ξ, and ∞. We skip writing the similar equations for the cases of the applied external fields in the y- and z-directions, where we only need to replace a in Eq. (3.10) with b and c, respectively.

Equations (3.17) and (3.18) are used for the analytical calculation of magnetic potentials produced by magnetic disks by assuming $a = b > c$. But, as mentioned earlier, the inconsistency in magnetic pole distribution due to the serial connection of the disks needs to be answered. Towards this goal, the model is further modified, and the magnetic thin film is approximated as a continuous line charge spanning the perimeter of the magnetic tracks. In order to confirm the inclusion of the shape-induced demagnetizing factors, the two models (i.e., the line charge model and the ellipsoidal model) are matched. To calculate the magnitude of the line charge density, results from Eqs. (3.17) and (3.18) are plugged into Eq. (3.6). The high aspect ratio of the magnetic thin films allows us to assume that the magnetic poles are confined to the edges and approximated as a line charge with a density expressed by

$$\lambda = \iota c \delta, \tag{3.20}$$

where ι is a numerical matching coefficient and is used to match the far field of the line charge model with the ellipsoid model with a good approximation. This approximation can be done by using the least squares analysis to minimum the difference between the volume averaged fields of the line charge from Eq. (3.20) and the analytical solution of Eqs. (3.17) and (3.18). For example, for a magnetic disk with a diameter and thickness of 20 μm and 100 nm, respectively, and for a linear strip with length and width of 3 μm, ι is 0.124 and 0.0406, respectively [6].

3.3.2   Hybrid (Finite-Element-Method/Analytical) Solution

The semi-analytical method mentioned in the last section dramatically reduces the computation time. But it is based on various assumptions, which may lower the accuracy of the model. This accuracy is acceptable for some applications, but to achieve better results, here another method is discussed [12]. This method, which is based on finite element methods (FEM) simulations, is fast and more accurate.

In this approach, FEM analysis is used to compute the magnetization of the magnetic thin films exposed to a uniform magnetic field along the track axis (x-direction). Then, the simulation is repeated for the fields applied along the other two axes [12]. This part of the analysis can be done, for example, by COMSOL or Ansys Maxwell software. Then, Eq. (3.6) is used to evaluate the equivalent magnetic charge distribution on the magnetic thin films. If we assume the magnetic thin films to be in the shape of connected disks, for example, the charge density on the axial surfaces and the disk perimeter are depicted by δ_s and δ_p, respectively. Their geometric variations can be modeled by hyperbolic and periodic functions [12]:

$$\delta_s(r, \theta) = \delta_{0s} \sin h\left(\vartheta \frac{r}{R}\right) \cos(\theta), \tag{3.21}$$

$$\delta_p(\theta) = \delta_{0p} \cos(\theta), \tag{3.22}$$

where $\sigma_{0p} = 6.2 H_{\text{ext}}$, $\sigma_{0s} = 0.12 H_{\text{ext}}$, and $\vartheta = 4$ are the optimal fitting parameters. Now, Eqs. (3.7) and (3.8) are used to calculate the magnetic potential and magnetic field intensity based on the calculated charge distribution. To show the agreement between this method and the full FEM-based simulations, the energy line cross sections based on them are plotted in [12]. Note that, as mentioned, the used resources in running simulations based on the method explained here are much less than running a full FEM analysis for analyzing the particle trajectories.

3.4 Particle Trajectory Analysis

To predict the particle trajectories, we consider the in-plane forces. Considering no acceleration, based on Newton's law for static force balance, $\sum \vec{F} = ma = 0$, Eq. (3.23) can be written for the in-plane forces.

$$\sum \vec{F} = \vec{F}_{\text{mag}} + \vec{F}_{\text{drag}} + \vec{F}_f + \vec{F}_{\text{th}} = 0, \tag{3.23}$$

where $F_{\text{mag}}, F_f, F_{\text{th}}$, and F_{drag} stand for the magnetic forces, friction forces, thermal forces, and drag forces, respectively. Note that in Eq. (3.23), the vertical forces (e.g., the gravity force) are not included. The thermal forces for the microparticles in the chips are weaker than the magnetic forces and can be neglected [13]. The weak friction (including rotational friction) forces are negligible. Also, since the magnetophoretic chips are typically coated with non-fouling layers, we can neglect the chance of particle-surface bond formation. Moreover, since in the magnetophoretic circuits, the particle trajectory is normally outside the magnetic patterns, considering the dynamics of particles moving over the 100-nm-thick magnetic thin film is not needed [14]. Hence, after canceling the negligible terms in Eq. (3.23), we realize that the drag force needs to be balanced by the applied magnetic force. On the other hand, for small spherical particles in aqueous media, using an overdamped first-order motion equation, the drag force can be written as $F_{\mathrm{drag}} = 6\pi \vec{v} \eta_f r_p$. Thus, by replacing the drag force with magnetic force, the particle velocity is expressed as Eq. (3.24) [15].

$$\vec{v} = \frac{\vec{F}_{\mathrm{mag}}}{6\pi \eta_f r_p}. \tag{3.24}$$

where η_f and r_p are the fluid viscosity and the particle radius, respectively.

Now, by plugging the magnetic force (from Eq. [3.4]) into Eq. (3.24), we can set up an equation of motion for the magnetic particles in a space-time varying magnetic field, which gives us the trajectory of the particles. This can be achieved by using a simple forward difference scheme, based on Euler finite time difference method, where the particle position at each time point, based on its previous position, is expressed by Eq. (3.25).

$$\vec{r}_i = \vec{r}_{i-1} + \vec{v}_{i-1}\Delta t, \tag{3.25}$$

where Δt is the time step [16]. Choosing a suitable time step is very important. Too small and too large time steps result in increased computation times and problems in the convergence of the solution, respectively. In other words, by choosing too small time steps, we force the system to divide the trajectory paths into too small sub-paths and compute them all. Also, by choosing very large time steps, the simulations would not be able to identify small trajectories, which results in unacceptable low-resolution results.

To conclude, the theoretical background needed for designing magnetophoretic circuits and analyzing their operation was provided. The discussion started with investigating the different types of magnetic materials. Then, the theory required for calculating the magnetic energies and forces was explained. The computational approaches including the semi-analytical method and the hybrid method were discussed. At the end of this chapter, a simple method for predicting the particle trajectory was explained.

References

1. Spaldin, N. A. (2010). *Magnetic materials*. Cambridge University Press.
2. Linderoth, S., & Khanna, S. N. (1992). Superparamagnetic behaviour of ferromagnetic transition metal clusters. *Journal of Magnetism and Magnetic Materials, 104*, 1574–1576.
3. Gregersen, E. (2011). *The Britannica guide to electricity and magnetism*. Britannica Educational Pub.
4. Poole, C. P., et al. (2014). *Superconductivity*. Elsevier Science.
5. Moskowitz, B. M. (1991). *Hitchhiker's Guide to magnetism*. http://www.irm.umn.edu/hg2m/hg2m_index.html

6. Lim, B., et al. (2014). Magnetophoretic circuits for digital control of single particles and cells. *Nature Communications, 5*, 3846.
7. Liu, J. P., et al. (2009). *Nanoscale magnetic materials and applications* (p. 1). Springer Science+Business Media, LLC (online resource).
8. Abedini-Nassab, R. (2019). Magnetomizcrofluidic platforms for organizing arrays of single-particles and particle-Pairs. *Journal of Microelectromechanical Systems, 28*(4), 732–738.
9. Stratton, J. A. (1941). *Electromagnetic theory*. McGraw-Hill Book Company, Inc
10. Chang, H. (1961). Fields external to open-structure magnetic devices represented by ellipsoid or spheroid. *British Journal of Applied Physics, 12*(4), 160.
11. Tejedor, M., et al. (1995). External fields created by uniformly magnetized ellipsoids and spheroids. *IEEE Transactions on Magnetics, 31*(1), 830–836.
12. Abedini-Nassab, R., et al. (2015). Characterizing the switching thresholds of magnetophoretic transistors. *Advanced Materials, 27*(40), 6176–6180.
13. Panigrahi, P. K. (2016). *Transport phenomena in microfluidic systems*. Wiley.
14. Abedini-Nassab, R., & Shourabi, R. (2022). High-throughput precise particle transport at single-particle resolution in a three-dimensional magnetic field for highly sensitive bio-detection. *Science and Reports, 12*(1), 6380.
15. Furlani, E. P., & Sahoo, Y. (2006). Analytical model for the magnetic field and force in a magnetophoretic microsystem. *Journal of Physics D: Applied Physics, 39*(9), 1724–1732.
16. Abedini-Nassab, R., et al. (2014). Optimization of magnetic switches for single particle and cell transport. *Journal of Applied Physics, 115*(24), 244509.

Chapter 4
Experimental Methods

In this chapter, the typical experimental methods used for fabricating magnetophoretic chips are discussed. I will explain the ones based on microfabrication
techniques. Although some other methods may also be used, the introduced methods
here are today considered widely acceptable and reliable approaches. For the fabrication process of the magnetomicrofluidic chips made of silicon and glass, please
see Sect. 7.3.2 Silicon-Glass-Based Magnetomicrofluidic Fabrication Protocol. In
addition to the chip fabrication methods, in this chapter, I provide the required information for packing methods, magnetic field stage, the control/monitoring system,
data collection, data analysis methods, and cell labeling techniques.

4.1 Microfabrication

Semiconductor industries have used microfabrication and nanofabrication techniques
for decades. These technologies and the required protocols are now well developed.
Since these technologies are suitable for fabricating the chips with micrometer or
nanometer patterns, researchers chose to use the same techniques for fabricating
microfluidic chips.

The idea is to use photolithography techniques to transport the required patterns
from the masks to the photoresist materials (which are light-sensitive materials that
typically are in liquid form initially and then solidify) on the chips. Then, these
patterned photoresist materials are used as guides to form other materials (e.g.,
metallic thin films, microfluidic channels, etc.). In the next subsections, a detailed
explanation for each step is provided. Here the fabrication steps, mostly taken from
[1], are explained in detail. Also, the schematic in Fig. 4.3 illustrates the fabrication
protocol.

© The Author(s), under exclusive license to Springer Nature Singapore Pte Ltd. 2023   59
R. Abedini-Nassab, *Magnetomicrofluidic Circuits for Single-Bioparticle Transport*,
https://doi.org/10.1007/978-981-99-1702-0_4

4.1.1 Photolithography

The process starts with choosing the substrate, which may be a silicon or glass wafer. The substrate needs to be cleaned thoroughly with acetone and isopropanol. Piranha solution (i.e., a mixture of sulfuric acid (H_2SO_4) and hydrogen peroxide (H_2O_2)) with a typical ratio of 3:1 can be used to clean organic residues off the chips. Since mixing the piranha solution is an extremely exothermic process, its handling requires extra care. After piranha cleaning, the chips are thoroughly rinsed with water thoroughly.

For fabricating various layers in the microfluidic chips, photolithography is used to pattern photoresist materials. Two types of photoresists, namely positive and negative photoresists, are commercially available. Positive photoresists are photo-sensitive materials that degrade when exposed to light and then are dissolved away in the photoresist developer. Thus, in the end, it leaves behind a photoresist coating where the mask was covering it. Negative photoresists in light exposure are polymerized or cross-linked and so solidify. Then, in the photoresist developer, the regions covered with the mask are removed. When photolithography is used to form metallic layers, a negative photoresist is suggested.

Here, some typical values for timings and temperatures are provided [1], which may vary based on the manufacturer's protocol. To perform the photolithography process using a negative photoresist, the photoresist is spin-coated onto the chip surface for ~5 s at 500 rpm, followed by ~30 s at 3000 rpm. A pre-bake step is performed at 90 °C for 120 s in an oven or 60 s on a hotplate. Then, coated chips are exposed to ultraviolet (UV) light for illumination time, illumination power, and wavelength of 12 s, 13.5 mW, and 365 nm, respectively. Next, the chips are baked (*i.e.*, post-exposure) at 90 °C for 120 s in the oven (or for 60 s on a hotplate). Then, the chips are placed in the photoresist developer at room temperature for 60 s. Finally, the chips are rinsed with deionized water and dried with nitrogen gas.

When using a positive photoresist, usually an adhesion promoter (e.g., P-20 [Shin-Etsu MicroSi, Pheonix, AZ]) is spin-coated on the chips before spinning the photoresist on them. All steps are similar to the case of negative photoresist, except the bake timings are different.

4.1.2 Metallic Thin Film Deposition

To form patterned thin films on the chips, after lithography, silicon chips are placed into a metal evaporator or sputtering system. Before it, the chips are suggested to be cleaned by plasma ashing (O_2) for ~60 s at 100 mW. Metal evaporator systems may use electron beams (e-beam) or electrical resistive heating techniques for evaporating the target metal. Also, sputtering systems use bombardment with energetic ions to release atoms from the target and pass into the gas phase. For example, using electron-beam evaporation, at an operating pressure of 1×10^{-5}, ~5 nm/100 nm stacks of Ti/

$Ni_{80}Fe_{20}$ or Ti/Au, at the rates of 0.2 nm/sec are deposited on the chips. The Ti thin film is used as an adhesion layer for better stability of the permalloy or gold thin films.

4.1.3 Lift-Off

After coating the whole wafer with metals to shape them with the lithographically patterned photoresists, they are placed in photoresist remover (e.g., NMP 1165), at 65 °C, for 5 min overnight. In this process, called lift-off, the photoresist and the overlaid metals on them are removed, leaving the patterned metallic layer. If needed, ~30 s of ultrasonication and rinsing by acetone and isopropanol can remove the unwanted residues.

4.1.4 Insulating Thin Film Deposition

In magnetophoretic circuits, different metallic layers exist. Also, on top of the chip, the particles and cells are exposed to the chip in a liquid (e.g., cell culture media or PBS). Thus, insulting layers both between the metallic thin films and on top of the chip are needed. Silicon dioxide is normally a good candidate. To create a 250 nm thick layer of SiO_2 as the insulating layer, Plasma Enhanced Chemical Vapor Deposition (PECVD) can be used. The deposition is done at 250 °C, at the rate of 35 nm/Sec.

Alternatively, SiO_2 films can be replaced by SU8, which shows better insulating properties in aqueous environments. A ~300 nm thick layer of SU8 photoresist can be created by mixing SU8 3005 with cyclopentanone (Microchem, Westborough, MA) at a 5:2 ratio. The resulting SU8 is spin-coated on the chips for 5 s at 500 rpm, followed by 30 s at 3000 rpm. The thickness of the SU8 layer can be controlled by adjusting the spin speed and, if required, the mixture ratio. After spinning, the wafers are baked on a hotplate at two steps of 65 °C for 30 s and 95 °C for 180 s. Then, the chips are exposed to UV light for 10 s at an illumination power of 13.5 mW at a wavelength of 365 nm. A post-bake step with timings and temperatures similar to the ones of the pre-bake step is needed. Finally, the chips are kept in SU8 developer for 180 s, rinsed with IPA, and dried with nitrogen. A photomask can be used to keep the electrical electrodes uncovered. But in the case of SiO_2 coating, the electrodes need to be uncovered with the reactive ion etching method, which is explained in the next subsections.

4.1.5 Soft Lithography

The microfluidic channels in the magnetophoretic chips can be based on either PDMS or Silicon. The PDMS-based microfluidic chips are normally fabricated using SU8 as a mold. To fabricate a SU8 mold, different SU8 types (e.g., 3005, 3015, or 3025) with different spin-coating speeds can be used to produce different channel depths. These values are reported in the manufacturer's protocol manual (Microchem). The SU8 lithography process is similar to the one explained in the previous subsection. After creating the SU8 mold, PDMS is poured on it and cured to solidify. Then, the PDMS slab with the desired microchannel design on it is peeled off. Now, to seal the microchannels, the PDMS and a glass slide are plasma ashed (O_2) for 45 s at 45 W and immediately stacked with pressure. The plasma ashing step is used for surface activation (*i.e.*, cleaning the surfaces and generating reactive chemical groups for covalent bonding). More information about PDMS bonding methods can be found in the available review papers [2]. The microfluidic inlet and outlets are created by punching holes in the PDMS slab.

Aligning the PDMS piece with micrometer patterns on magnetophoretic circuits may be challenging. Hence, as an alternative method, it is possible to directly use patterned SU8 as the chip to be sealed with a PDMS piece. The sealing process can be based on two different methods.

In the first method, the SU8 top layer is activated (*i.e.*, hydroxyl groups are generated on its surface) [3] by dipping the chip into sulfuric acid (96%) for 8 s. The chip is then carefully rinsed with water and dried with nitrogen. At the same time, the PDMS piece is plasma ashed, and then, immediately, the two pieces are aligned, brought in contact, and heated at 120 °C for more than 30 min.

In the second method, an extra ~200 nm SiO_2 layer can be created on the chip, which then can be plasma activated together with a PDMS piece. The generated Si-OH groups in the ashing process form relatively strong Si-O-Si bonds after placing the two pieces in contact.

4.1.6 Reactive Ion Etching

To make the electrical contacts with the electrode pins on the chip, they need to be accessible. Hence, if the chip is covered by SiO_2 (e.g., for insulation purposes, as explained in the previous subsections), the electrode surfaces need to be uncovered. Towards this goal, a positive photoresist is photolithographically patterned to cover the whole chip, except the electrodes. Then, reactive ion etching (RIE) is used to remove the extra silicon dioxide covering the electrodes. Finally, the chips are cleaned with acetone and IPA and then dried with nitrogen gas.

4.2 Surface Functionalization

In order to provide the right surface for the particles to move and also to make a good environment for the cells to behave normally, different chip surface functionalization is needed. Here, some functionalization methods needed in the magnetophoretic circuits are discussed.

4.2.1 SiO_2 Surface Passivation

In order to prevent the chip surface from forming bonds with the particles, it is covered with a non-fouling layer. The chip passivation can be done with a Poly(oligo(ethylene glycol) methyl ether methacrylate) (POEGMA) brush layer. POEGMA, a derivative of the widely used polymer poly(ethylene glycol) (PEG), with its 3-dimensional "bottle-brush" structure which offers a higher oligo(ethylene glycol) functional group densities at the solid/water interface, shows fantastic non-fouling properties compared to the linear PEG [4].

First, the chips are immersed in a 10% solution of 3-aminopropyltriethoxysilane (Gelest, Inc.; Morrisville, PA) in ethanol for 3 h, followed by rinsing with ethanol and deionized water. The chips are dried in an oven for 1 h at 120 °C. Then, they are left at room temperature to cool down. Next, the chips are put in a dichloromethane solution containing 1% α-bromoisobutyryl bromide and 1% triethylamine (Sigma Aldrich; St. Louis, MO) for 30 min. After rinsing in fresh dichloromethane, ethanol, and deionized water, the chips are centrifuged and then oven-dried. Then, under an argon environment, the chips are immersed in a degassed solution composed of 350 mL deionized water, 23 mg copper (II) bromide, 33 mg bipyridyl, 600 mg ascorbic acid, and 55 g of inhibitor-free poly(ethylene glycol) methyl ether methacrylate (Mn ≈ 300) and left to stir for 6 h. Finally, after rinsing with deionized water and being dried, the chips are ready [1, 5].

The polymerization quality on various metal oxide surfaces is studied [6]. PECVD can be replaced with atomic layer deposition (ALD) to achieve good-quality thin films of SiO_2, TiO_2, ZrO_2, and Al_2O_3. The surfaces with and without POEGMA coatings are exposed to a 1 mg/mL solution of Cy5-BSA in 1X PBS buffer for 4 h and then rinsed with a PBS solution containing 0.1% Tween 20. Then, the chips are centrifuged at ~5000 rpm for 15 s to wick away excess liquid and then dried. Figure 4.1 illustrates the fluorescent images of the samples with absorbed proteins. This figure shows POEGMA coating nicely prevents protein adsorption.

Figure 4.2 shows the POEGMA growth on the various substrates. The POEGMA brush thickness is increased over time and reaches 50–70 nm. These consistent results show reliable film growth to thicknesses well above the minimum required values for POEGMA surfaces to present non-fouling behavior. Figure 4.2b depicts

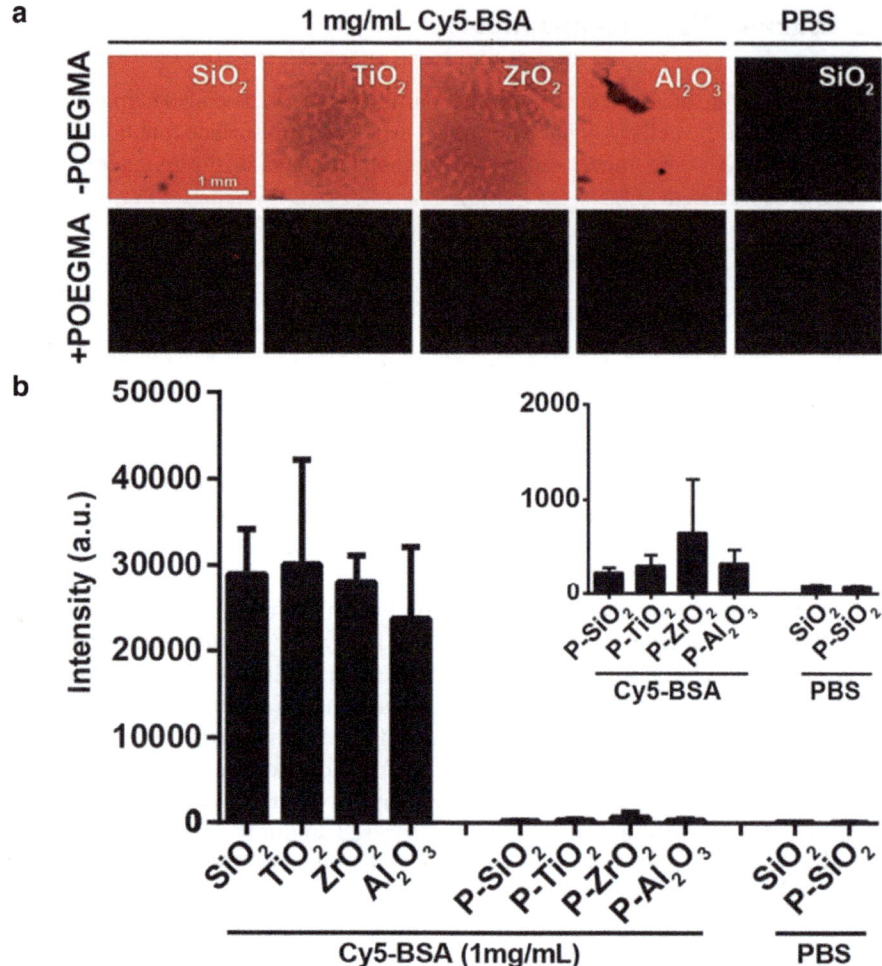

Fig. 4.1 Protein adsorption on various oxide surfaces with and without POEGMA is shown. (**a**) Fluorescence images of Cy5-BSA adsorption are illustrated. Pure SiO$_2$ surfaces and the ones coated with POEGMA both treated with PBS are also shown as control experiments. (**b**) Fluorescence intensities received from residual Cy5-BSA bound to surfaces are presented. A zoomed view of low fluorescence intensity data is shown in the inset. Reprinted with permission from [6]. Copyright 2017 American Chemical Society

that POEGMA does not grow on chips coated with (3-Aminopropyl) triethoxysilane (APTES) alone. It shows that POEGMA only initiates from a surface with an immobilized bromide initiator [6].

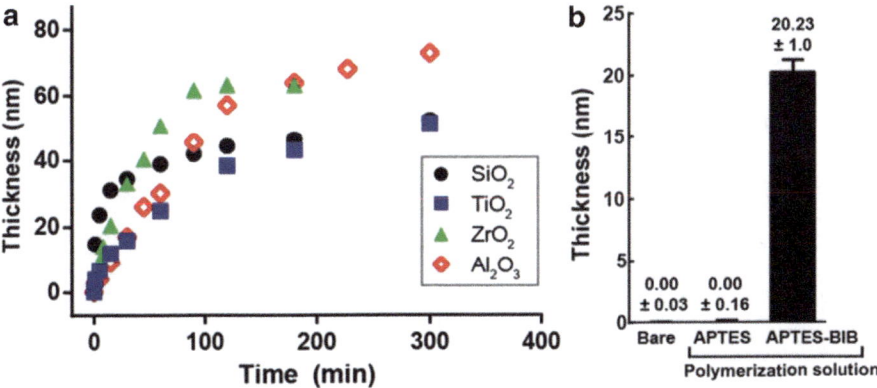

Fig. 4.2 POEGMA brush growth by SI-ATRP on different surfaces is shown. (**a**) POEGMA brush thickness is measured. (**b**) A control experiment that shows surface-immobilized bromide initiator is needed for POEGMA brush growth is illustrated. Reprinted with permission from [6]. Copyright 2017 American Chemical Society

4.2.2 SU8 Surface Passivation

POEGMA can also be grown on top of the SU8 layer by surface-initiated atom transfer radical polymerization in the SU8-based chips [7]. Towards this goal, the SU8 top layer is activated and hydroxyl groups are created on it. This process is explained in detail elsewhere [3]. Then they are conjugated to APTES. In this process, amino groups of APTES react with the epoxy groups of the SU8. Then, the chip is ready to grow POEGMA brushes on it.

A fabrication procedure schematic is illustrated in Fig. 4.3. In this figure, SU8 is used as the insulating layer. Panels j–l depict the fabrication steps when a microfluidic channel is required, while panels n–q show the steps for a chip without microfluidic channels. Panel m illustrates the grown POEGMA on the chip.

4.2.3 Surface Functionalization for Biocompatibility Purposes

To help the cells behave normally after sorting on the magnetophoretic chips, the chip surface is functionalized with different biomaterials. Here, the functionalization steps are explained, and in the next chapters, their biocompatibility effects are studied.

One of the materials commonly used in cell culture is Poly-lysine. 5 mg of the poly-lysine is solved in deionized water. After cleaning and drying the chips, they are coated with the prepared solution (1 mL/25 cm^2) and incubated for 5 min. Then, the chips are rinsed with deionized water and dried.

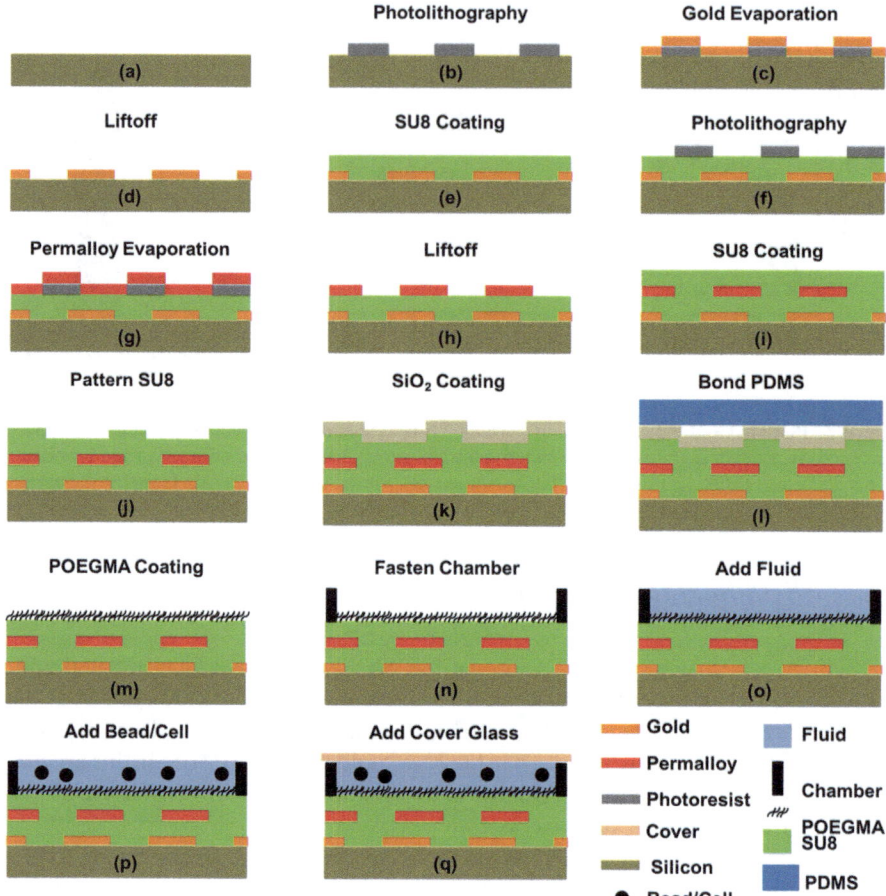

Fig. 4.3 Chip fabrication steps are shown. (**a**) Starting from a cleaned silicon wafer, (**b**) a photoresist is patterned. (**c**) Then a thin stack of Ti/Au is evaporated onto the entire surface, (**d**) after which lift-off is performed, and finally, (**e**) a SU8 layer is applied to act as an insulator (**e**). (**f–i**) Steps (**b–e**) are repeated with the magnetic permalloy layer. (**j**) In the fabrication of the chips with microfluidic channels, another SU8 layer is patterned and aligned on the chip. (**k**) It is coated by SiO$_2$, and then the (**l**) microfluidic channels are sealed with PDMS. In the chips without channels, (**m**) the SU8 layer is coated with POEGMA brush, (**n**) and then a 3D printed chamber is installed on it. (**o**) DI water, PBS, or cell culture media, and then (**p**) cells or beads are added. Finally, (**q**) a planar viewing window is achieved by covering the chip with a coverslip, as opposed to the PDMS shown in (**l**)

The next material is a protein called collagen. For example, bovine collagen protein (e.g., from ThermoFisher Scientific) can be diluted to 100 μg/mL in 20 mM acetic acid. After cleaning and drying the chips, the chip surface is covered with the prepared solution at 4 °C for one hour. Then the chips are rinsed with sterile 1X PBS and gently dried.

To coat the chips with gelatin, it is added to 50 mL of DI water. The solution is then placed on a hotplate at 90 °C for 3 min. After filtering the solution, cleaning the chips, and drying them, the solution is added to the chip surface. The chips are kept at room temperature for 60 minutes. Finally, the chips are rinsed with sterile 1X PBS and gently dried with nitrogen gas.

Fibronectin, a protein found in the extracellular matrix, is widely used in cell cultures and artificial tissues. To coat the magnetophoretic chips, fibronectin is diluted in a sterile-balanced salt solution to 100 ug/mL. The chips are cleaned, dried, and then covered with the prepared fibronectin solution. After incubating them at room temperature for 1 h, they are briefly rinsed with PBS and gently dried with nitrogen gas.

4.3 Packing Methods

4.3.1 3D Printing

The magnetophoretic circuits can be combined with PDMS-based or silicon-based microfluidic channels, the fabrication method of which is explained above. But, with the development of 3D printers, it is also possible to, alternatively, print a chamber and stick it on top of the chip. A glass coverslip can be used as a window on top to seal the chamber. The sealed chamber prevents evaporative drying and facilitates microscopic imaging (see Figs. 4.3 and 4.5).

4.3.2 Electrical Connections

The electrodes on the magnetophoretic chips, which provide the electrical signals to the transistor gates in the circuits, need electrical connections to the outside world. Toward this goal, two methods can be used. The first method is based on the wire bonding technique, where electrical wires with diameters of ~50 μm are bonded to the electrodes on the chip. The other end of the microwires can be connected to a supply board, which provides the required electrical signals. Alternatively, it is possible to use commercially available connectors or clips.

4.4 Rotating Magnetic Field Generator

The driving force for the particles on magnetophoretic circuits is normally provided by a rotating magnetic field. This field can be created by various coils surrounding the chip. In the magnetophoretic circuits operating in a two-dimensional field, four coils do the job, while in the circuits operating in a three-axial field, five coils are needed. An iron plate machined into a four-pole structure, in which each arm is wrapped with magnet wire (20 AWG) can be used. When needed, an extra coil underneath the chip makes the vertical magnetic field. The apparatus needs to be powered by programmable power supplies (see Fig. 4.5).

4.5   Monitoring System

An input/output control board can be used to monitor the signals from the chip and send the control signals to the chip. The monitoring signals can be the ones received from the temperature or gas sensors, and the control signals are the ones sent to the gate of the transistors to control the particle trajectory. Also, software prepared in LabVIEW or other platforms works with the control board.

A picture of a customized LabVIEW code for the mentioned control system is shown in Fig. 4.4. This software reads the data from a temperature sensor (see Fig. 4.5e), based on which controls the electrical current passing a resistive heater.

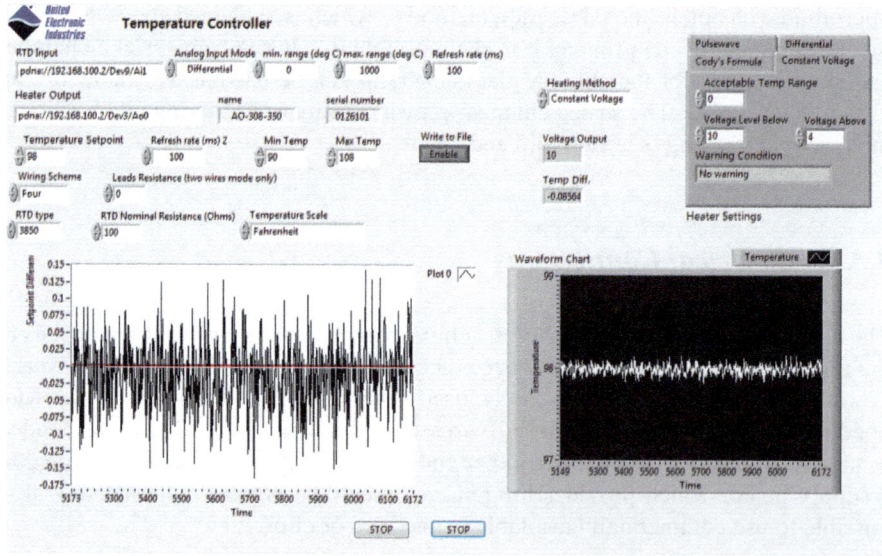

Fig. 4.4 A sample temperature controller LabVIEW front panel is illustrated. The code is written by Elijah Weinreb, from Duke University. It is used to send control signals to a heater to adjust the chip temperature at 98 °F (~37 °C)

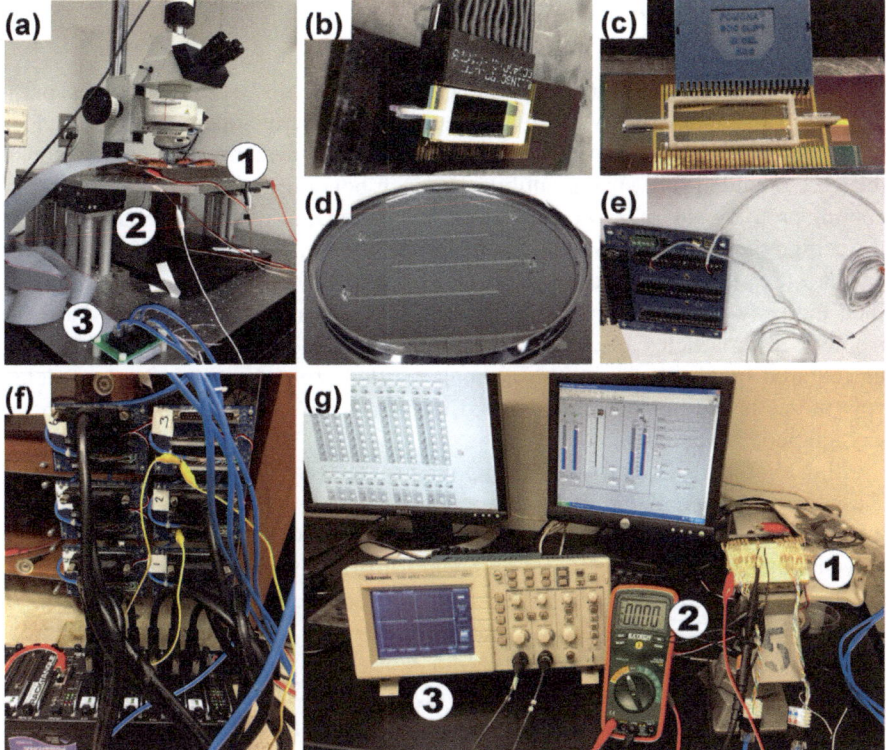

Fig. 4.5 Experimental setup is presented. (**a**) The imaging station and magnetic field stage are shown. The rotating magnetic field is created by four coils (1) with the fifth coil sitting underneath the platform (2). The magnetic fields and gate electrical currents were controlled with a custom-designed electrical board (3). The chip can be mounted in (**b**) a PCI connector or (**c**) an IC test clip. (**d**) A microfluidic-based chip is illustrated. (**e**) An I/O card connected to two temperature sensors is shown. (**f**) The control system includes the I/O cards installed on a rack. (**g**) The test setup is shown. The voltage and current of electrical resistors (1) are measured by a multimeter (2) and an oscillator (3)

4.6   Image Processing

A video camera mounted on a microscope is needed to record movies and obtain the particle positions and trajectories. An image processing software or code then can be used to detect the particles in the obtained images.

4.7   Cell Magnetic Labeling

To be manipulated by the magnetophoretic circuits, the cells or other biological particles need to be magnetically labeled. To do so, magnetic nanoparticles are provided
to the cells to uptake them. Alternatively, commercially available magnetic nanoparticles are bounded (e.g., using antibody-antigen bounding) to the cell surface. The
labeling protocol is typically provided by the manufacturer. But it normally consists
of washing the cells in PBS with centrifuging and re-suspending in FBS-PBS. The
cells are then ready for adding the commercial selection cocktail and incubation.
Then the cells are washed again in PBS and re-suspended in FBS-PBS. Now the
magnetic nanoparticles are added, mixed, and incubated to give them time to make
the required bonds. A final wash normally is done to remove the extra particles.
Now the cells are magnetically labeled and are ready to be transported with the
magnetophoretic circuits.

All in all, the methods for fabricating the magnetophoretic chips were explained.
These techniques are the microfabrication techniques widely used in semiconductor
processing. Typically, to form a thin film on a silicon wafer, the process starts with
conventional lithography followed by metal lift-off. Various instruments, including
metal evaporators, PECVD, and ALD, can be used to coat the chip with a layer of
metal, oxide, etc. The microfluidic channels are widely fabricated using soft lithography. This method has also been used for fabricating magnetomicrofluidic chips;
however, introducing the microchannels into silicon wafers is considered a better
method which is explained in the next chapters. Surface functionalization is an
important step in making the magnetomicrofluidic chips ready for bio-applications.
The packing methods, magnetic stage, control system, and data collection methods
were discussed. Also, it was explained how the cells are labeled before entering
magnetomicrofluidic chips.

References

1. Abedini-Nassab, R., et al. (2015). Characterizing the switching thresholds of magnetophoretic
   transistors. *Advanced Materials, 27*(40), 6176–6180.
2. Borók, A., Laboda, K., & Bonyár, A. (2021). PDMS bonding technologies for microfluidic
   applications: A review. *Biosensors, 11*(8), 292.
3. Tao, S. L., Popat, K., & Desai, T. A. (2006). Off-wafer fabrication and surface modification of
   asymmetric 3D SU-8 microparticles. *Nature Protocols, 1*(6), 3153–3158.
4. Hucknall, A., Rangarajan, S., & Chilkoti, A. (2009). In pursuit of zero: Polymer brushes that
   resist the adsorption of proteins. *Advanced Materials, 21*(23), 2441–2446.
5. Hucknall, A., et al. (2009). Simple fabrication of antibody microarrays on nonfouling polymer
   brushes with femtomolar sensitivity for protein analytes in serum and blood. *Advanced Materials,
   21*(19), 1968–1971.

6. Joh, D. Y., et al. (2017). Poly(oligo(ethylene glycol) methyl ether methacrylate) brushes on high-κ metal oxide dielectric surfaces for bioelectrical environments. *ACS Applied Materials & Interfaces, 9*(6), 5522–5529.
7. Abedini-Nassab, R., et al. (2016). Magnetophoretic transistors in a tri-axial magnetic field. *Lab on a Chip, 16*(21), 4181–4188.

Chapter 5
Magnetophoretic Circuits Operating in an In-Plane Magnetic Field

Two main types of magnetophoretic circuits are available. The first type discussed in this chapter is the one that needs a two-dimensional (2D) in-plane magnetic field to operate. The second type, discussed in the next chapter, operates in a tri-axial (i.e., three-dimensional [3D]) field.

In this section, magnetophoretic circuit elements for transporting the particles (e.g., conductors and diodes), storing them (e.g., capacitors), and switching their trajectory (e.g., transistors) are introduced.

The goal of designing magnetophoretic circuits is to provide an energy distribution with energy wells that move smoothly along the magnetic tracks in our desired direction. Hence, by connecting linearly magnetizable circular magnetic patterns (i.e., disks or half-disks) in series, a periodic asymmetry in magnetic thin film patterns is introduced. In an external rotating magnetic field, the magnetization of the magnetic disks attempts to synchronize with its direction. This assumption, in the case of permalloy films, which do not show a strong hysteresis behavior, is reasonably acceptable.

From classical magnetostatics [1], when linear magnetizable materials are used the dipolar energy minima form where the outward normal component of the magnetic film pattern curvature is parallel to the external magnetic field. Thus, when a magnetic disk is exposed to a rotating magnetic, two energy wells form on their opposite sides (i.e., the north and south poles). This is shown in the simulation results in Fig. 3.2. Magnetic particles and cells close to the magnetic disks tend to move to these energy wells. In a rotating magnetic field, the energy wells continuously move around the disk perimeter. In serially connected magnetic disks (see Fig. 3.2), the particles periodically switch between the north and south poles of the adjacent magnetic disks and move exactly two disk periods, D, along the magnetic tracks at each rotating field cycle. In the magnetic track design, a linear segment is considered to connect the two adjacent magnetic disks and provide an energy barrier that prevents the local energy minima from switching between the magnetic tracksides (see Fig. 3.2). The strip must be wide enough to provide the required energy barrier able to compete against the energy well produced by the magnetic disks. But too wide a strip is

R. Abedini-Nassab, *Magnetomicrofluidic Circuits for Single-Bioparticle Transport*, https://doi.org/10.1007/978-981-99-1702-0_5

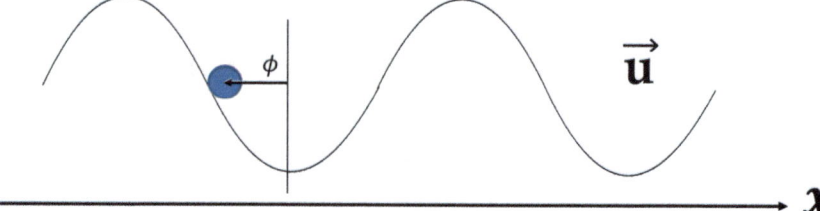

Fig. 5.1 A traveling periodic energy landscape is illustrated. A particle, shown in blue, is carried by a traveling energy landscape, with a phase lag of ϕ. The speed of the energy landscape is u at $+$ x direction

also problematic. Particle transport between two adjacent magnetic disks requires overlapping their energy wells at their intersection, which is prohibited by too wide strips. A sample experimentally detected particle trajectory is shown in Fig. 3.2b.

Assume a periodic energy landscape that moves constantly (see Fig. 5.1), and carries a particle at the energy minima with position ϕ in the traveling reference frame. This energy can be expressed as Eq. (5.1).

$$U = U_0 \sin(Kx - \omega t), \tag{5.1}$$

where U_0, t, $K = 2\pi/\lambda_w$, and $\omega = 2\pi f$ are the energy amplitude, time, wave number, and angular frequency, respectively, while λ_w and f stand for the wavelength and frequency, respectively. The force acting on the particle can be derived by the negative gradient of the energy and can be written as Eq. (5.2).

$$\vec{F} = -KU_0 \cos(Kx - \omega t). \tag{5.2}$$

Also, the particle position is written as

$$\phi = Kx - \omega t \tag{5.3}$$

We have

$$\frac{d\phi}{dt} = K\vec{u} - \omega. \tag{5.4}$$

By inserting Eqs. (3.24) and (5.2) into Eq. (5.4), it can be expressed as

$$\frac{d\phi}{dt} = \frac{-K^2 U_0 \cos\phi}{6\pi \eta_f r_p} - \omega, \tag{5.5}$$

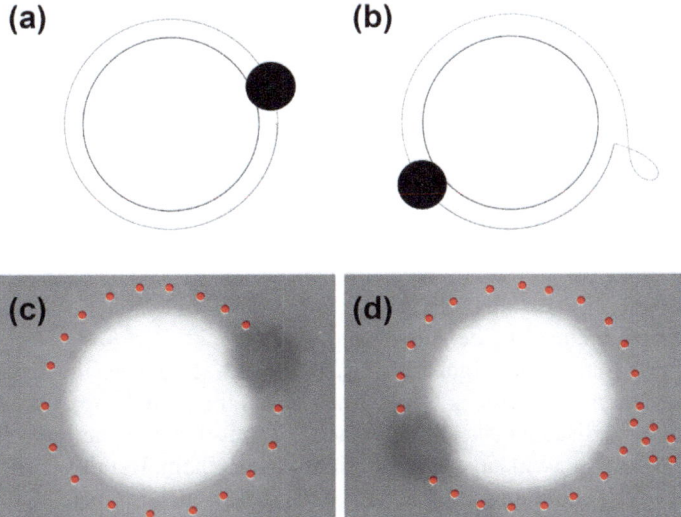

Fig. 5.2 Particle trajectories based on simulations for the **a** phase-locked and **b** phase-slipping regimes and the corresponding experimental observations for the **c** phase-locked and **d** phase-slipping regimes are illustrated. The figure is reprinted from [2] with permission from the Royal Society of Chemistry

Equation (5.5) can be written in the form of an equation of a non-uniform oscillator depicted in Eq. (5.6).

$$\frac{d\phi}{dt} = -(\omega + \omega_0 \cos\phi). \tag{5.6}$$

In Eq. (5.6), for $\omega < \omega_0$ regime, there exists a ϕ, such that $\frac{d\phi}{dt} = 0$. This condition is called the phase-locked regime, where the particles follow the energy landscape with a phase lag of ϕ. In this regime, by increasing the frequency, the phase lag increases. Figure 5.2a, c illustrate the theoretical and experimental particle trajectory in this regime. But for $\omega > \omega_0$, no ϕ can be found, such that $\frac{d\phi}{dt} = 0$. This condition is called the phase-slipping regime, where the viscous force is beyond a critical threshold, and the particles cannot follow the external field. In this regime, the particles periodically slip out of a potential energy well and are captured by the adjacent energy well. This trajectory is shown in Fig. 5.2b, d.

Now, before entering the discussion about the magnetophoretic circuit elements, similar to the electronic materials, in which the movement of electrons is investigated, here I will introduce the same concept for the movement of particles in engineered materials. Drawing inspiration from electric materials, magnetometamaterials with tunable magnetic matter conductivity are introduced [3]. These materials transport magnetic materials, as opposed to electrons. They can be tuned to operate as an insulator (i.e., do not transport the particles), semiconductor (i.e., transport some of the particles), and conductors (i.e., transport all the particles). Magnetometamaterials are composed of magnetic disks fabricated on a silicon chip with a gap of a specific size in between them.

When magnetic disks are assembled in a row, an analysis similar to what is explained above can be performed. Note that here the disks are separate and no magnetic strip connects them. At frequencies lower than a critical frequency, f_{cm}, the particles have enough time to follow the energy wells and rotate around a single disk. The energy landscape distribution for this case is shown in Fig. 5.3a. Also, the particle trajectories based on both simulations and experiments are illustrated in Fig. 5.4a, e. This case is called the "insulating regime," in which the particles are not transported and the material behaves as an insulator.

At operating frequencies higher than f_{cm}, the phase lag between the particle and the applied magnetic field becomes larger. Thus, when the particle approaches the area between the two adjacent disks, it is distanced from its leading energy well, which is labeled 1 in Fig. 5.3b, with an energy barrier between them. But this particle is close to the energy well circulating the next magnetic disk, which is labeled 2 in Fig. 5.3b. Thus, the particle is transported to the next magnetic disk. The particle trajectories based on simulations and experiments for this case are depicted in Fig. 5.4b, f, respectively.

But f_{cm} is a function of many factors such as particle size, particle magnetization, and gap size. For example, a large gap between the magnetic disks, as shown in Fig. 5.3c, leads to an energy barrier between the particle and the next magnetic disk. This energy barrier prevents the particle from being transported to that disk. Thus, because of the random fluctuations in the mentioned parameters, some particles move from one disk to the next one and ultimately proceed along the magnetic tracks while others do not. This regime is called the "semiconducting regime," where the conductivity of the proposed material can be tuned by adjusting the applied frequency. By lowering the frequencies, the particles follow the energy wells rotating the magnetic disk close to them, and the conductivity is lowered. By increasing the frequency, the conductivity is enhanced.

By further increasing the frequency, two different conducting regimes are observed. It was mentioned that increasing the applied frequencies results in higher conductivities. Hence, at frequencies higher than the ones of the semiconductor regime, all the magnetic particles move to the next magnetic disks and move along the magnetometamaterial. This condition is called the "non-slipping conducting regime." At even higher frequencies (i.e., $f > f_{cs1}$), the system enters the slip-out regime explained above, and the particles slip out of the given energy well. For the case of a single magnet, the particle trajectory was analyzed above (see Fig. 5.2). But in the case of having multiple magnetic disks in a row (i.e., in magnetometamaterials), when the particle slips out, it is captured by the energy well circulating the next magnetic disks. Figure 5.3d shows the energy landscape in this case. Figure 5.4c and g depict the particle trajectory simulation prediction and the experimental trajectory results, respectively. This case is called the "slipping conducting regime."

At even higher frequencies (i.e., $f > f_{cs2}$), the particles slip out more frequently, and before the particle approaches the next magnet, it experiences a curvy trajectory.

Fig. 5.3 The energy landscapes for the magnetometamaterial at different operating regimes are illustrated. **a** At frequencies lower than f_{cm}, the particle rotates around a single magnetic disk. The dashed black line stands for the particle trajectory. **b** At frequencies higher than f_{cm}, the particle moves to the next disk. The dashed black and red lines depict particle trajectories around the first disk and the second disk, respectively. **c** A large gap between the magnetic disks leads to an insulating regime. The dashed black line depicts the particle trajectory. **d** At frequencies above f_{cs}, the particle moves to the next magnetic disk. The particle, as opposed to a circular slip-out path (dashed black line), experiences the dashed red line trajectory. Numbers 1 and 2 are the leading energy well on the first disk and the next disk, respectively. The circular arrow shows the direction of the external field rotation. In these simulations, the disk and particle diameters are considered to be 10 μm and 2.8 μm, respectively. The figure is reprinted with permission from [3] Roozbeh Abedini-Nassab, Physical Review Applied, 17, 014,020, 2022. Copyright (2022) by the American Physical Society. https://doi.org/10.1103/PhysRevApplied.17.014020

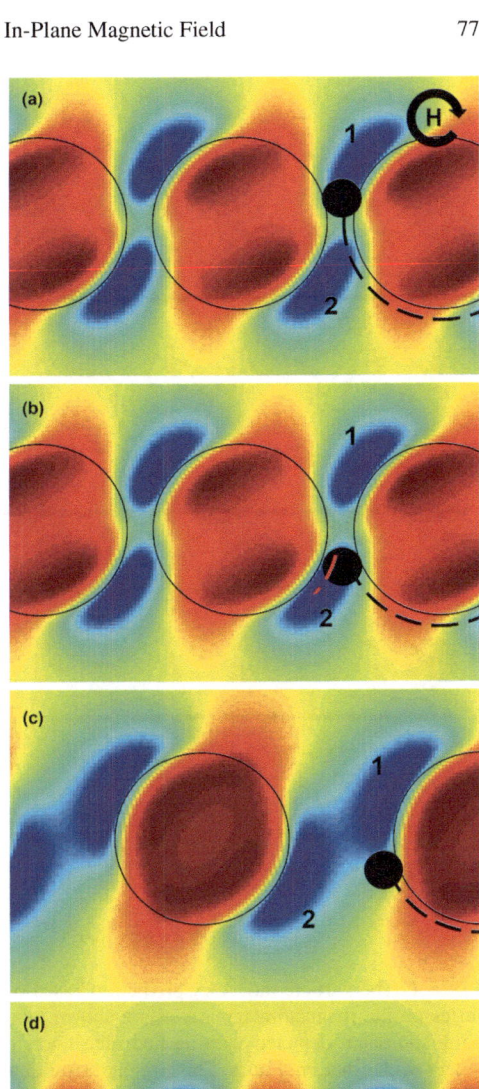

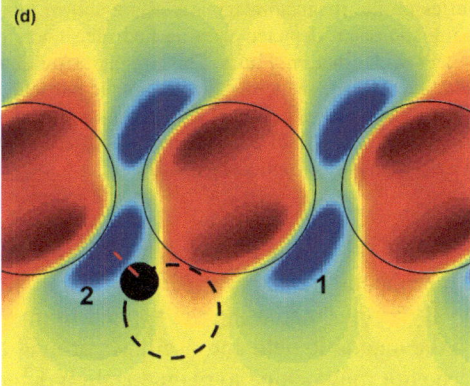

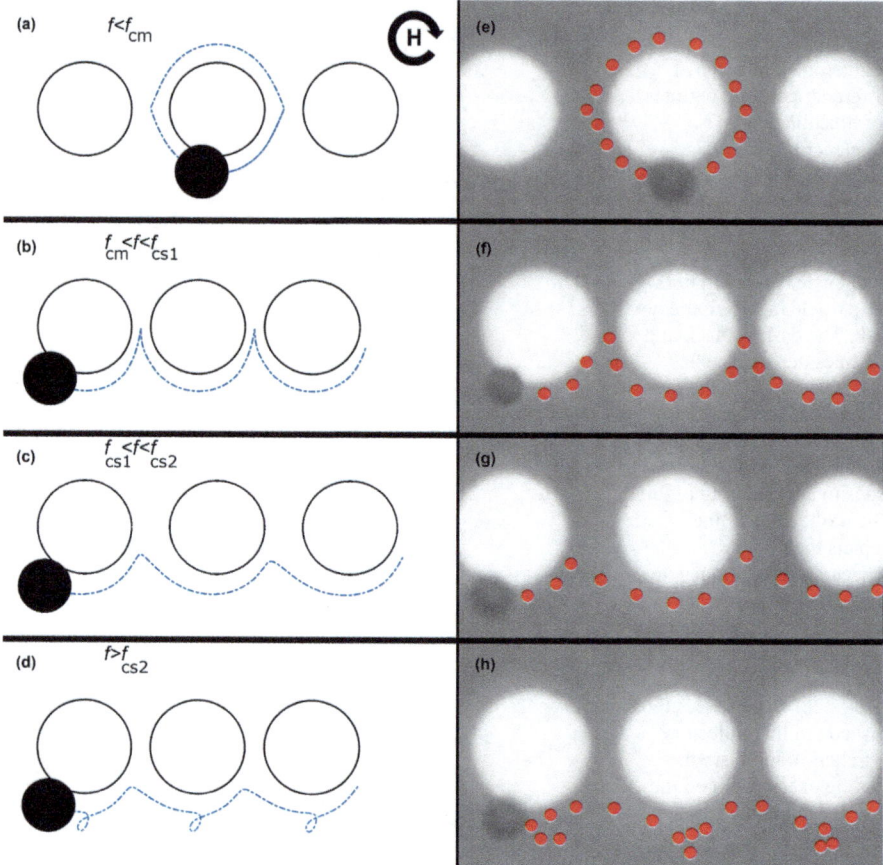

Fig. 5.4 The particle trajectory prediction based on simulations and the real experimental trajectories on the magnetometamaterial are shown. **a–d** The particle trajectory predictions based on simulations are depicted with dashed blue lines. **e–h** The experimental particle trajectories are shown with dotted red lines. The circular arrow shows the direction of the external magnetic field rotation. The magnetic disk and magnetic particle mean diameters are 6 μm and 2.8 μm, respectively. The figure is reprinted with permission from [3] Roozbeh Abedini-Nassab, Physical Review Applied, 17, 014,020, 2022. Copyright (2022) by the American Physical Society. https://doi.org/10.1103/PhysRevApplied.17.014020

Figure 5.4d and h illustrate the simulation particle trajectory prediction and the experimental trajectories, for this regime, respectively.

By further increasing the applied frequency ($f > f_{cs3}$), the particle cannot answer to the magnetic field and oscillates around a fixed point on the disk perimeter. This behavior may be due to magnetization defects or imperfect geometries in the fabricated magnetic disks. In this regime, the particles are trapped on a pinning site. At these frequencies, the system enters a regime called the "deep insulating regime," in which the magnetometamaterial behaves similarly to an insulator material.

Fig. 5.5 The gap size vs. frequency phase diagrams for the magnetometamaterial are plotted. The solid and dashed lines depict the boundaries of the conducting and insulating regimes, respectively. The area between the two lines shows the semiconducting regime. The magnetic disk diameters are 6 μm, and the mean diameters of the particles are (a) 2.8 μm and (b) 5.7 μm. The figure is reprinted with permission from [3] Roozbeh Abedini-Nassab, Physical Review Applied, 17, 014,020, 2022. Copyright (2022) by the American Physical Society. https://doi.org/10.1103/PhysRevApplied.17.014020

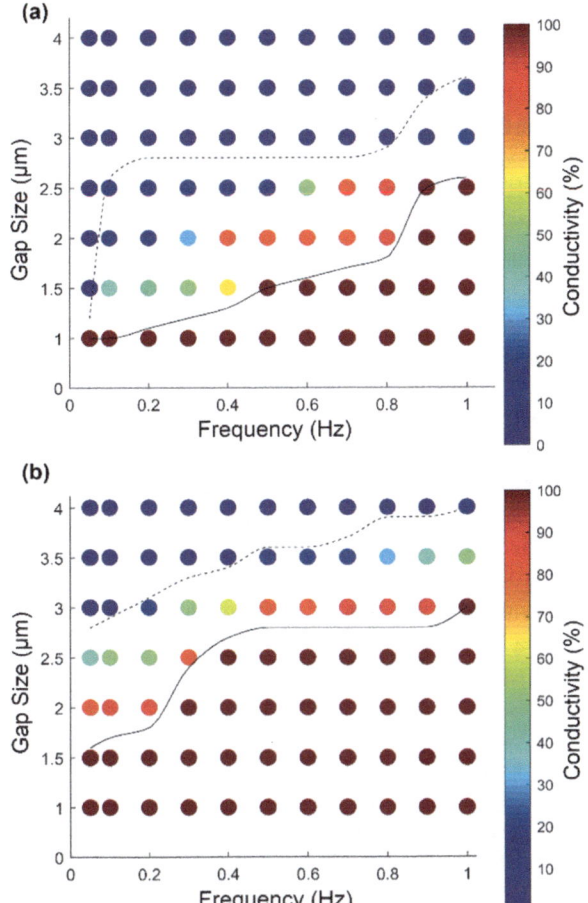

To show the effect of the gap size and the applied frequency, the phase diagrams in Fig. 5.5 are plotted. These plots show the percentage of the particles being transported along the magnetic tracks on the magnetometamaterial (i.e., conductivity). The solid line and the dotted line in this figure show the boundaries for the conducting regime and the insulating regime, respectively. The region between these two lines depicts the semiconducting regime. Figure 5.5a and b represent the results for magnetic particles with mean diameters of 2.8 μm and 5.7 μm, respectively.

In addition to the 1D material introduced in the previous section, in which the magnetic disks are arranged in a single row, two-dimensional magnetometamaterials are also introduced. But in an in-plane rotating field, the particles experience closed-loop trajectories. It is shown in the next subchapters that these magnetometamaterials can transport the particles in a tri-axial magnetic field.

5.1 Conductors

When the external magnetic field frequency is sufficiently low, as discussed above, the particles can follow the energy wells. Thus, at these frequencies, the particles are synched to the external field and move along the magnetic tracks at a speed that is linearly proportional to the applied frequency.

$$u = \kappa \mathrm{f}, \tag{5.7}$$

where κ is the mobility of the particles. Equation (5.7) describes the particle velocity and its linear dependence on the external stimuli. This equation is similar to Ohm's law in conductors in electrical circuits, which shows a linear relationship between the velocity of the electrons and the applied external electric field (i.e., $\overrightarrow{u} = \kappa \overrightarrow{E}$, where E is the electric field). Although more complicated forms of both relations can be written, I have considered their simplest form.

In a magnetic field rotating clockwise, the particle illustrated in Fig. 5.6 moves from left to right. Now in a counterclockwise magnetic field, the particle moves from right to left. That means we can control the particle trajectory direction by adjusting the external field, and the design can transport the particles bidirectionally. This is similar to the behavior of the conductors in the electrical circuits, which conduct electrical currents bidirectionally.

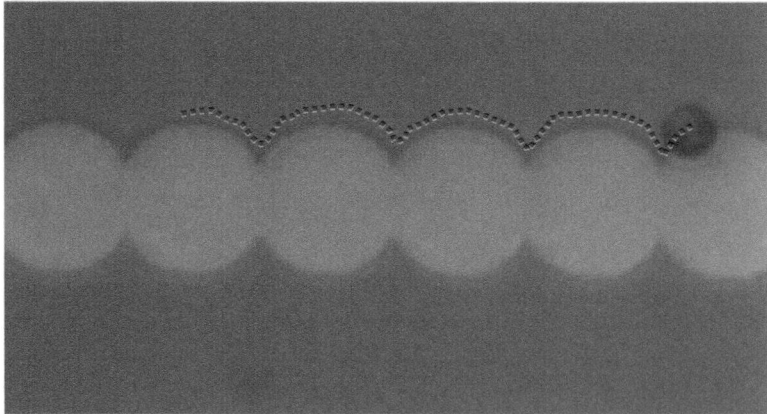

Fig. 5.6 A magnetically labeled human CD4+ T cell moves on a magnetophoretic conductor. A clockwise in-plane rotating magnetic field is applied. The blue dotted line shows the cell trajectory

5.2 Diodes

A magnetophoretic diode (i.e., rectifier) is designed by introducing an additional asymmetry to the conducting paths. For example, forming a T-junction with two magnetic tracks, composed of half-disks, as shown in Fig. 5.7, can be considered. Similar to the electrical diodes, the magnetophoretic diode may be biased in two modes. In the forward bias (i.e., in the clockwise magnetic field in Fig. 5.7a–e), the particle on the horizontal track moves from left to right. When the particle approaches the T-junction, in an external magnetic field parallel to the vertical magnetic track, two separate energy wells form on the opposite sides of the strip (see Fig. 5.7c). By further rotating the external field (see Fig. 5.7d), the two energy wells merge and form a single well. This energy well eliminates the energy barrier on the strip. Also, it is deeper on the right side of the vertical track. Thus, it moves the particle from the left to the right of the magnetic strip. However, in the reverse bias (i.e., in the counterclockwise magnetic field in Fig. 5.7f–j), the particle approaches the T-junction from the right. At the T-junction, always deeper energy well is seen on its right side, and thus the particle does not pass the vertical strip. Hence, the approaching particle follows the energy well moving along the vertical magnetic track. This particle trajectory results in a zero net current to the left side of the T-junction. In other words, the magnetophoretic diode conducts the magnetic particles in the forward bias; however, it does not transport them in the reverse bias.

5.3 Capacitors

In order to study the particles of interest, we need to keep them on a spot. When a short magnetic track (i.e., a track composed of one or two connected magnetic disk(s)) is exposed to an external rotating magnetic field, the adjacent particles circulate in a closed loop (see Fig. 5.8). This behavior is similar to one of the capacitors in the electrical circuits storing electrons.

Another magnetophoretic capacitor is designed based on the diode with a closed magnetic path, as illustrated in Fig. 5.9. The T-junction entry is similar to the diode design in the forward bias mode that allows particles or cells to enter the capacitor. The entrapped particles move in a closed loop inside the capacitor. The diode entry does not allow the particles to exit the capacitor even in the reversed magnetic field rotation. Loading the capacitor with B- and T-lymphocyte cell pairs are also shown in Fig. 5.9c–e.

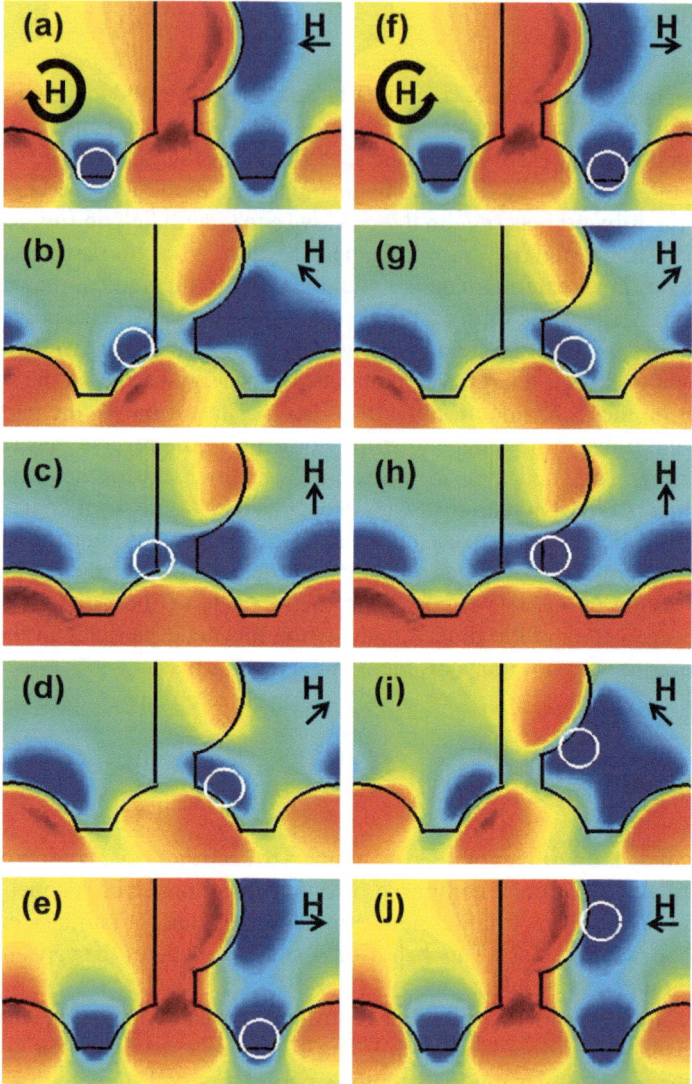

Fig. 5.7 The potential energy simulation of the magnetophoretic diode is illustrated. The potential energy distribution for the diode in **a–e** forward bias, where the external magnetic field rotates clockwise, and **e–h** reverse bias, where the external magnetic field rotates counterclockwise, are presented. The black arrows and small circles denote the external magnetic field direction and the particles, respectively

Fig. 5.8 A magnetized
human CD4+ T cell stored in
a capacitor is shown. The
blue dotted line shows the
cell trajectory

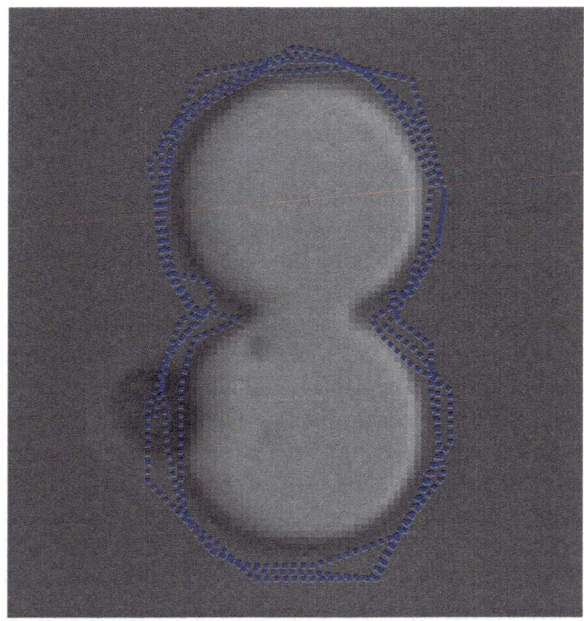

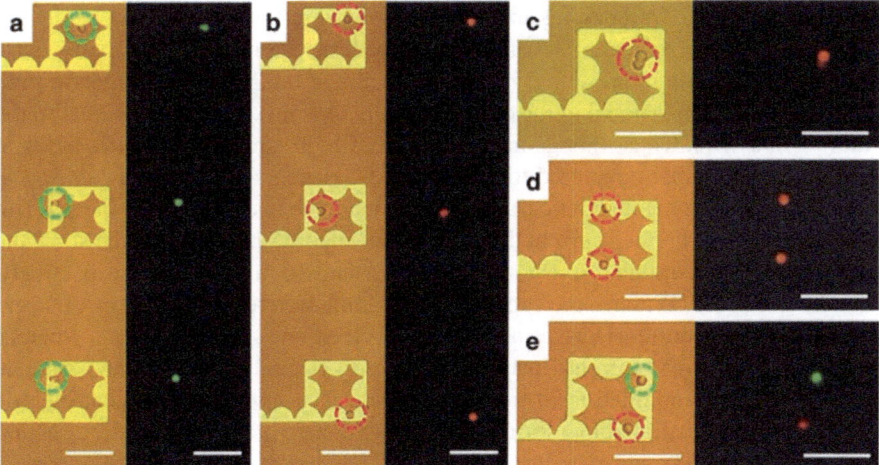

Fig. 5.9 Bright-field and fluorescent images of **a** single T lymphocytes and **b** B lymphocytes loaded in capacitors are illustrated. **c** Cell mitosis in a capacitor is shown. **d** A pair of B cells is shown. **e** A pair of B- and T cells is formed in the capacitor. Scale bars: 40 μm. The figure is taken with permission from [4] under a Creative Commons Attribution 3.0 Unported License. http://creativec ommons.org/licenses/by/3.0/

5.4 Transistors

In the magnetophoretic circuits, in order to precisely control the particle trajectory, it is needed to switch their trajectory path at specific spots. This task is done by magnetophoretic transistors. Similar to the electronic transistors that turn on/off by an applied gate control signal, we introduce a current-carrying metallic thin film as the gate to control the transistor particle transport. Let's first check whether a current-carrying metallic thin film (i.e., microwire) crossing over a magnetic strip connecting the magnetic disks in a track can switch the particles from one side of that track to the other side. If so, we have found a method to inverse the particle transport direction [5]. Based on the simulation results presented in [5], an electric current passing through a metallic thin film of fixed width and perpendicular to the magnetic track can lower the depth of the local energy well; however, it cannot compete with the energy barrier produced by the magnetic strip. A rotated, widened, or laterally shifted microwire cannot solve the problem either. In [5], to study the bistable energy well more accurately, a line section of the energy landscape along the wire is extracted from the simulations which clearly shows the bistable energy. Hence, no straight wire is suitable in designing a transistor for switching the particle trajectory in this magnetic geometry.

Proper operation of the proposed transistor needs the elimination of the bistable energy well. Hence, introducing an asymmetric current line may help in achieving particle switching. An example of the wire with an asymmetric geometry, wire tapered at both ends is presented in Fig. 5.10. Since the produced magnetic field with the wire is proportional to the current density, it is inversely proportional to the width of that wire. Thus, this geometry produces high-strength magnetic fields at the center of the taper compared to the two ends. Hence, a tapered wire shifted to the opposite side of the magnetic track can eliminate the bistable energy barrier. But note that sufficient applied current densities are needed. Figure 5.10a and b show the energy landscapes simulation results for 100 mA and 10 mA currents, respectively, at the height of 0.12D above the substrate. The simulation result for 100 mA current (sufficient electric current) has a single energy well on one side of the track, whereas the results for 10 mA current (insufficient electric current) still show the bistable energy well. The annihilation of the energy barrier by applying sufficient electric current is better illustrated by plotting the energy distribution along the line section of the wire axis illustrates (see Fig. 5.10c). For this geometry, at ~40 mA electrical current the barrier is destroyed. The tapered section of the wire in the simulations of Fig. 5.10 has a width of 0.08D and a length of 0.42D. The effect of the lateral position of the wire is also studied, the results of which are shown in Fig. 5.10d for a constant applied current. Based on this study, the optimal lateral position for the wire is at the center of the gap.

Parameters to be considered here are the tapered section geometry and the particle size. Wider tapered sections need higher switching currents to maintain the overall current density. Longer wires do not significantly affect the switching behavior. In Fig. 5.10e, the switching efficiency as a function of the particle size relative to the

Fig. 5.10 Simulation results for a transistor based on a whole disk track and a tapered current sheet are presented. The energy landscape in a horizontal plane at the height of 0.12D above the substrate is plotted. The applied current is **a** 100 mA and **b** 10 mA. **c** The potential energy distribution along the wire axis for electrical currents of 10 mA (black), 50 mA (red), and 100 mA (blue) is presented. **d** The potential energy distribution along the wire axis for 50 mA current, where the center of the taper is shifted in the lateral (x) direction by amounts of 0.00D, 0.08D, 0.16D, 0.25D, and 0.33D is presented. The blue arrow indicates the increasing lateral shift. **e** The transistor switching efficiencies for the applied currents of 0 to 20 mA for the geometry of (a) and (b) for particles with radii of 0.12D (dashed black), 0.16D (dashed red), 0.25D (dashed blue), 0.33D (solid black), 0.50D (solid red), 0.67D (solid blue), and 0.83D (solid green) are shown. The black arrows and the black dotted lines depict the 100 Oe external magnetic field and the particle trajectory, respectively. The disk diameter is considered to be 0.83D. The figure is taken from [5] with permission

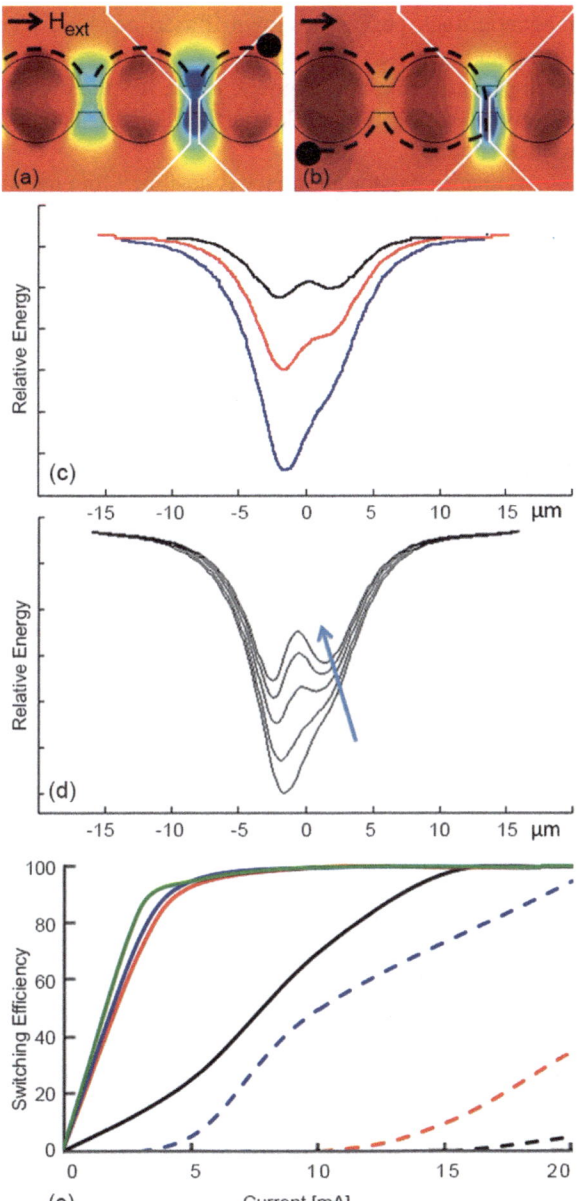

magnetic track period is shown. Based on these results, currents of 5–10 mA are sufficient for switching particles with diameters equal to or larger than the magnetic track period across the magnetic track. But the energy barrier made by the magnetic strip is too strong for smaller particles, and thus they need stronger electric currents to be switched.

Another transistor geometry is designed for enabling the magnetophoretic diodes to conduct particle currents in the reversed bias [4, 5]. This transistor becomes more important when it is used in the capacitors shown in Fig. 5.11, where the entrance is based on the diode design. In this case, the transistor is used to retrieve a particle or cell from the capacitor to, for example, perform follow-on genotypic analyses. This magnetophoretic transistor is designed by placing a diagonal microwire oriented at a 60° angle relative to the horizontal axis on the opposite side of the T-junction. As illustrated in Fig. 5.11, when a current of 100 mA is passing through the gate of the transistor, the bistable energy minima on the sides of the junction merge and form a single energy minimum on top of the junction (see Fig. 5.11b). Thus, at the switching time, the particle close to the microwire moves from the right of the T-junction to its left side and continues moving along the negative x direction of the horizontal track. By turning off the electrical current, the next particles approaching the junction cannot move to the left anymore (see Fig. 5.11b). In the example shown in Fig. 5.11, the deterministic nature of particle transport is used for synchronizing the switching task with the external rotating magnetic field clock to pass the second particle, but not the first and third particle, across the T-junction from right to the left [4].

To find the optimal gate geometry to minimize the required electrical currents for switching the particles, the switching efficiencies for diferrent orientations and positions of the wire are studied, the results of which are illustrated in Fig. 5.11 [5]. The simulation results show that sufficiently large electrical currents are needed to eliminate the energy barrier on the magnetic strip and transport the particles across the junction. The transistor switching efficiencies for different microwire angular orientations and for a particle with a diameter of 0.23D are studied in [5]. Since the goal is to shift the particles to the left side of the vertical magnetic strip in Fig. 5.11, the center of the wire is fixed at the left edge of the midpoint of this strip. In these calculations, the width of the wire, the diameters of the disks, and the lengths of the strips are 0.14D, 0.90D, and 0.10D, respectively. Based on the simulation results, microwires making an angle of $\alpha = 60°$ with the horizontal axis result in optimal switching efficiencies. So, for this transistor geometry and microwire orientation, the switching efficiency for transporting particles of different sizes is studied. The results indicate two important facts. First, the switching efficiency for the larger particles is higher. Also, compared to the other transistor types, the transistor introduced in Fig. 5.11 needs much larger currents (~60 mA) for appropriate operation.

One of the best transistor geometries is the one in which a gap is introduced in the magnetic track, and the transistor gate is placed on it (see Fig. 5.12) [4, 5]. This design behaves similarly to a semiconducting junction, which controls the particle current crossing the gap with the applied gate electrical current. A bistable energy distribution with minima at opposite sides of the gap is seen in the simulation results. Hence, particles approaching the gap do not cross the gap and stay on the same

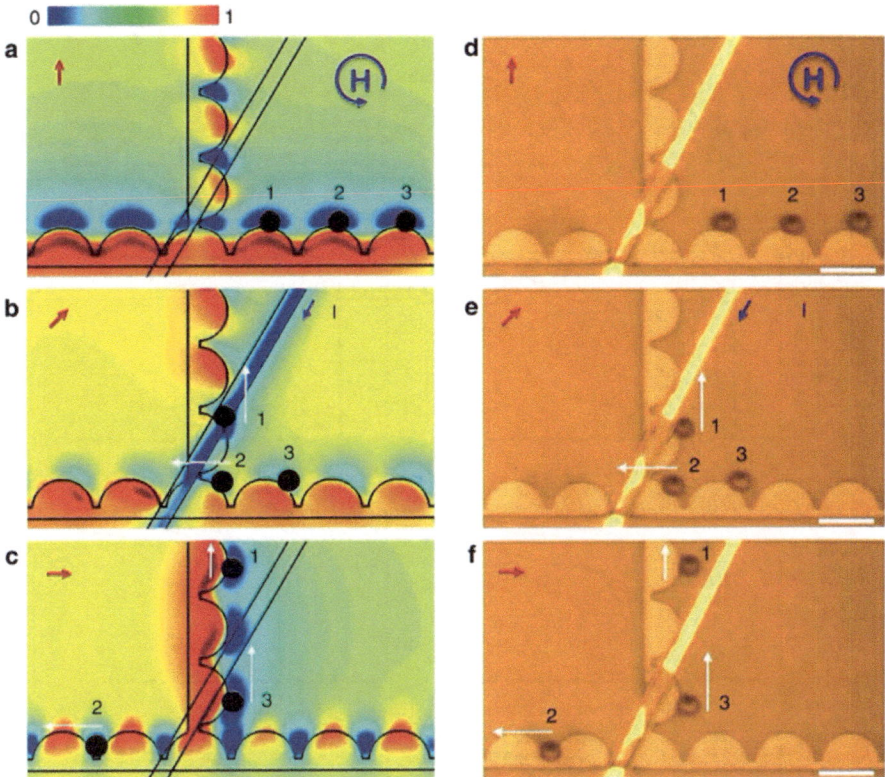

Fig. 5.11 A diagonal microwire is superimposed on the diode to selectively transport the single particles across the T-junction. Simulation results **a–c** and the experimental images **d–f** for sequential field angles (shown with the red arrows) with rotation steps of 45 degrees are shown. In panels a and d, the first particle is approaching the T-junction and the gate current is off. In panels b and e, the second particle is approaching the T-junction and the current is on. In panels c and f, the third particle is approaching the T-junction and the gate current is off. The white arrows represent the direction of the magnetic bead movement. Scale bars, 20 μm. The figure is taken with permission from [4] under a Creative Commons Attribution 3.0 Unported License. http://creativecommons. org/licenses/by/3.0/

magnetic track. A weak gate current also is not sufficient to change this situation and, as shown in Fig. 5.12a, the particles stay on the same side of the gap, resulting in particle movement direction reversal. But a sufficiently strong gate current annihilates the energy barrier and created a deeper energy well on the other side of the gap to attract the particle (see Fig. 5.12b). Thus, the particles cross the gap. Similar results are also seen for curved gates (Fig. 5.12c).

To understand the robustness of this switching mechanism, which can vary due to random fluctuations in the particle size and its magnetic susceptibility, in a non-dimensional analysis, the effect of the random parameters on transistor switching efficiency can be investigated. In this analysis, the dimensionless parameter $\beta =$

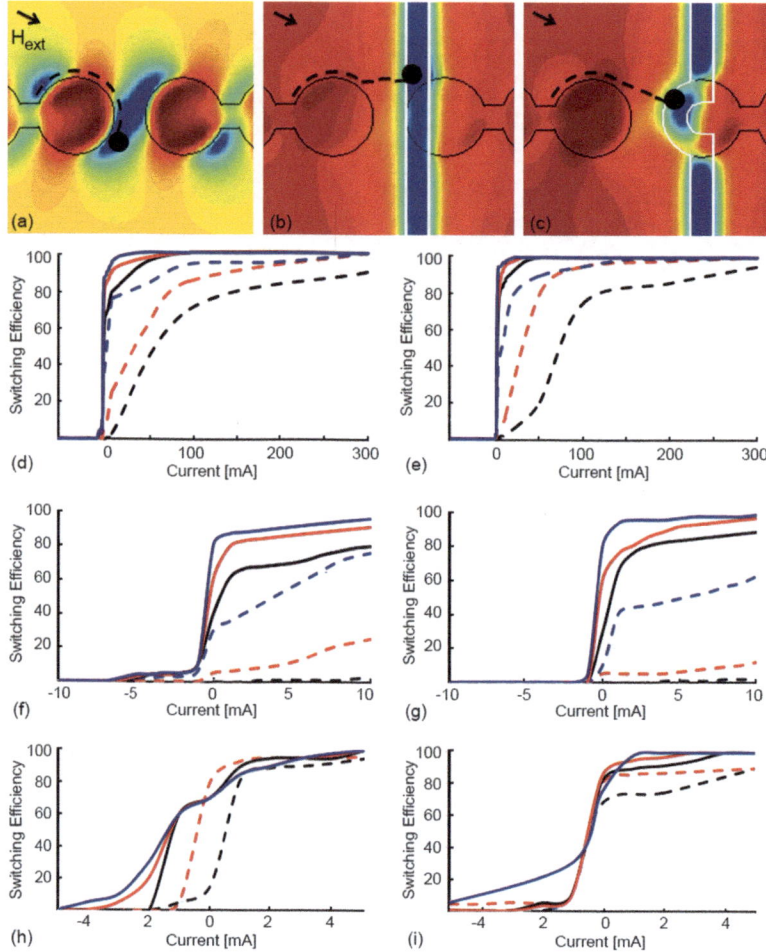

Fig. 5.12 Simulation results for a transistor composed of a whole-disk track with a semiconducting gap are shown. The energy distribution is presented in (a–c) as a function of the electrical current strength and geometries. The wire width is assumed to be fixed at 0.12D. **a** The energy landscape in the absence of a gate is shown. **b** The energy landscape for a transistor with a straight gate is shown. **c** The energy landscape for a transistor with a curved gate is shown. **d,f** The switching efficiencies for the transistor with a straight gate are plotted. **e,g** The switching efficiencies for the transistor with a curved gate are plotted. The plots in (f) and (g) are zoomed versions of the efficiencies illustrated in (d) and (e), respectively. In (d–g) plots for $\beta = 1.0$ (solid blue), $\beta = 0.9$ (solid red), $\beta = 0.8$ (solid black), $\beta = 0.7$ (dashed blue), $\beta = 0.6$ (dashed red), and $\beta = 0.5$ (dashed black) are illustrated, where the magnet diameter is 0.83D. **h** The switching efficiency for a transistor with a straight gate is presented. **i** The switching efficiency for a transistor with a curved gate is shown. In (h,i), $\beta = 1$, while the ratio of the particle radius to magnet radius is 0.15 (dashed black), 0.25 (dashed red), 0.35 (solid black), 0.45 (solid red), and 0.55 (solid blue). The black arrow in (a–d) stands for the direction of a 100 Oe external magnetic field. The blue and red areas denote the regions with low and high magnetic energies, respectively. The black dotted lines represent the particle trajectory. The figure is taken from [5] with permission

r_p/r_G, where r_p and r_G are the particle radius and the magnetic track gap size, is introduced. $\beta = 1$ corresponds to a particle radius equal to the gap size. Figure 5.12d and f depict the switching efficiency for the straight wire geometry of Fig. 5.12b for $0.5 < \beta < 1.0$, where Fig. 5.12f provides a zoomed-in view of the switching efficiency at low current. Similarly, Fig. 5.12e and g represent switching efficiency for the curved wire geometry of Fig. 5.12c for the same range of b values.

Based on the simulation results, the introduced transistors show higher switching efficiencies for the large particles (i.e., for large β). But too large particles can cross the gap even with a zero gate current. For example, for $\beta > 0.8$, more than 50% of the particles cross the gap, even when no gate current is applied. In this situation, an electrical gate bias current of -10 mA (i.e., a bias current in the opposite direction compared with the switching currents) is necessary to prevent this leakage of particles across the gap, called "particle leakage current."

The transistor with a curved gate wire, illustrated in Fig. 5.12, shows similar behavior, with some quantitative differences. The required negative bias electrical current to prevent the particle leakage current in this design is smaller (~ 2 mA) compared to the one mentioned above. Also, these transistors need slightly larger gate currents for switching the particles with high probabilities.

The effect of the size of the magnetic disks with respect to the size of the particles on the switching efficiency for a fixed particle to gap ratio is shown in Fig. 5.12. In particular, see panels h and i in this figure, where the switching efficiency for the straight and curved wire geometries, respectively, for $\beta = 1$, are plotted. In these simulations, the ratio of the particle diameter to the disk diameter varies between 0.15 and 0.55. The general trends indicate that the switching efficiency for particles with larger sizes relative to the disk diameter is better. But it must be noted that for large particles, a negative electrical bias gate is required to eliminate the particle leakage currents. Considering the two important mentioned facts about the effect of the particle size on switching efficiency, the optimal particle to magnet size ratio is approximately 0.5, which requires gate currents of ~ 65 mA for the straight or curved wires.

The position of the gate with respect to the gap plays an important role in the transistor switching efficiency [5]. Figure 5.13 presents the simulation results for a wire having a width of 0.25D, whose center is laterally shifted to four different positions relative to the gap center. These simulation results do not suggest placing the wire exactly at the center of the gap, as shown in Fig. 5.13a, because the resulting transistor does not operate well. The applied gate current is relatively ineffective both at pushing the particles across the gap and preventing particle leakage. To achieve better results, the gate needs to be shifted to one side of the gap. Shifting it by a distance equal to half the wire width, as illustrated in Fig. 5.13b, lowers the required negative bias to prevent the particle leakage current. It also enhances the transistor switching efficiency (see red curves in Fig. 5.13e–j). For example, in Fig. 5.13j, for $\beta = 1$ and particle mean radius of 0.20D, $+4$ mA and -2 mA gate currents are sufficient to turn the transistor on or off, respectively. Further shifting of the wire away from the central axis (see Fig. 5.13c and d) leads to an overall drop in the

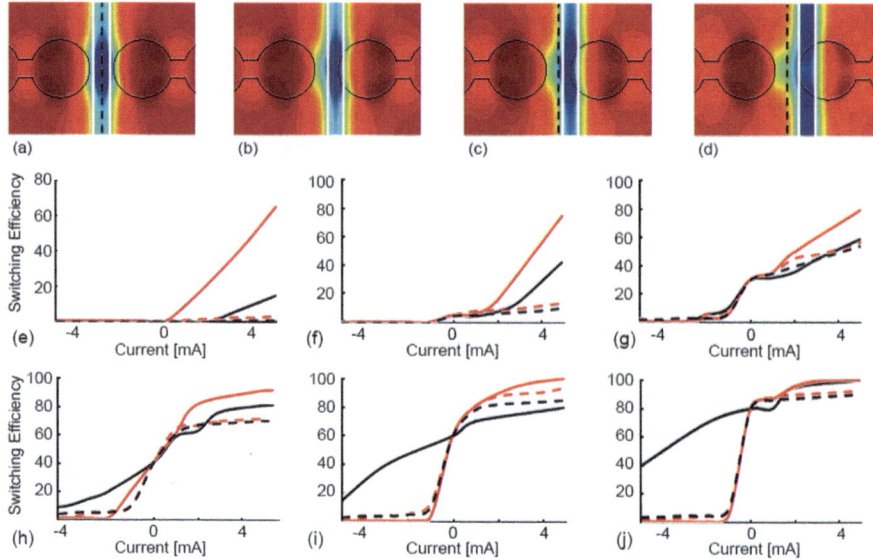

Fig. 5.13 Simulation results for studying the effect of the gate lateral position on the switching efficiency of the transistor composed of the whole-disk track with a semiconducting gap are shown. The magnetic energy landscape for a gate with a fixed width of 0.25D whose center is shifted away from the gap center by **a** 0D, **b** 0.13D, **c** 0.16D, and **d** 0.29D are shown. Switching efficiencies are plotted for each of these cases for **e** β = 0.5, **f** β = 0.6, **g** β = 0.7, **h** β = 0.8, **i** β = 0.9, and **j** β = 1. In (e–j), the curves correspond to the designs of parts (a) solid black, (b) solid red, (c) dashed red, and (d) dashed black. The ratio of the mean particle diameter to the disk diameter is 0.25. In parts (a–d), the black dashed line depicts the center of the gap. The figure is taken from [5] with permission

switching efficiency (see Fig. 5.13e–j). Thus, the mentioned gate position seems to be an optimal parameter to be considered in fabricating the introduced transistors.

The simulation results discussed above show that the transistors based on the whole-disk tracks with a semiconducting gap can achieve good performances with the lowest current switching thresholds. The ability to operate at small gate currents is important because it lowers the power consumption and the generated resistive heat. This parameter becomes even more important when thousands of transistors are integrated into a tiny chip.

The transistor with a semiconducting gap can operate in three different modes of (i) off, (ii) attractive, and (iii) repulsive [6], as shown in Fig. 5.14. The attractive mode is what we already discussed above, the energy simulation and experimental results of which are illustrated in Fig. 5.14b and e. As the simulation result in Fig. 5.14b and comparing it with the case with a zero gate current (see Fig. 5.14a) indicates, the gate electrical current eliminates the double well distribution and the energy barrier at the center of the gap and provides the required energy well over the gate to attract the particles. It is also confirmed in the experiments (see Fig. 5.14e). In this design, the gate is placed on the other side of the gap to attract the particle. In the repulsive mode (see Fig. 5.14c and f), sufficiently strong gate currents can annihilate the double well energy barrier seen in Fig. 5.14a (off mode) and induce the cell to transfer across the gap away from the wire. In this mode, the particle and the gate are on the same side of the gap, and the particle is repelled from the gate to the other side of the gap. The experimental particle trajectory in Fig. 5.14f confirms this finding. Thus, when the gate current is zero (i.e., the transistor is off), the particle approaching the gate experiences an inverse transport direction (see Fig. 5.14d). But with sufficient gate currents, including the attractive and repulsive modes, the approaching particle moves across the gap (see Fig. 5.14e, f).

The measured switching thresholds for the attractive and repulsive transistor modes are illustrated in Fig. 5.15. Based on the hypothesis that arose from simulation (i.e., the optimal transistor performance is achieved for the gap sizes similar to the particle diameter) and considering the typical size of human T cells (6–10 µm) as the test bioparticles, transistors with gap distances ranging from 5 to 15 µm are tested [7]. In the transistors with gaps smaller than 8 µm, many particles cross the

Fig. 5.14 The operational modes and energy landscape for the transistor composed of the whole-disk track with a semiconducting gap are shown. The magnetic potential energy distributions are presented in (a–c) with their corresponding experiments shown in (d–f), respectively. The energies at line cross sections (dotted lines) are illustrated **a** for off mode, **b** for the attractive mode, and **c** for the repulsive mode. The chip experiences a constant magnetic field along the positive x-direction shown by the black arrow. The field rotation sense is depicted by the circular arrow and the gate electrical currents are shown by the red arrows. The figure is taken from [7] with permission

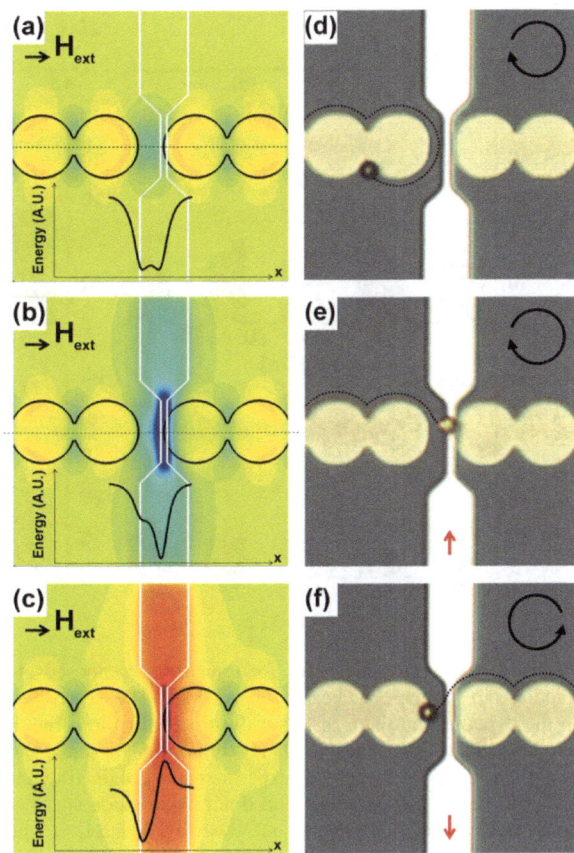

gap in the absence of a gate current. Thus, this geometry is not suitable for a magnetophoretic transistor. The transistors with gaps larger than 10 μm require excessively large gate currents for reliable switching. Thus, this geometry is not suitable either. Hence, gap dimensions of 8 or 10 μm, which are slightly larger than the average cell diameter, seem to be appropriate choices.

Two approaches of "dynamic" and "static" transistor tests for evaluating its performance are considered. The dynamic tests are more representative of the device operating conditions, compared to the static tests. The parameter L is defined as the distance of the left edge of the wire from the disk edge on the opposing side of the gap. Assuming the transistor geometries of d = 10 μm and L = 9 μm, the

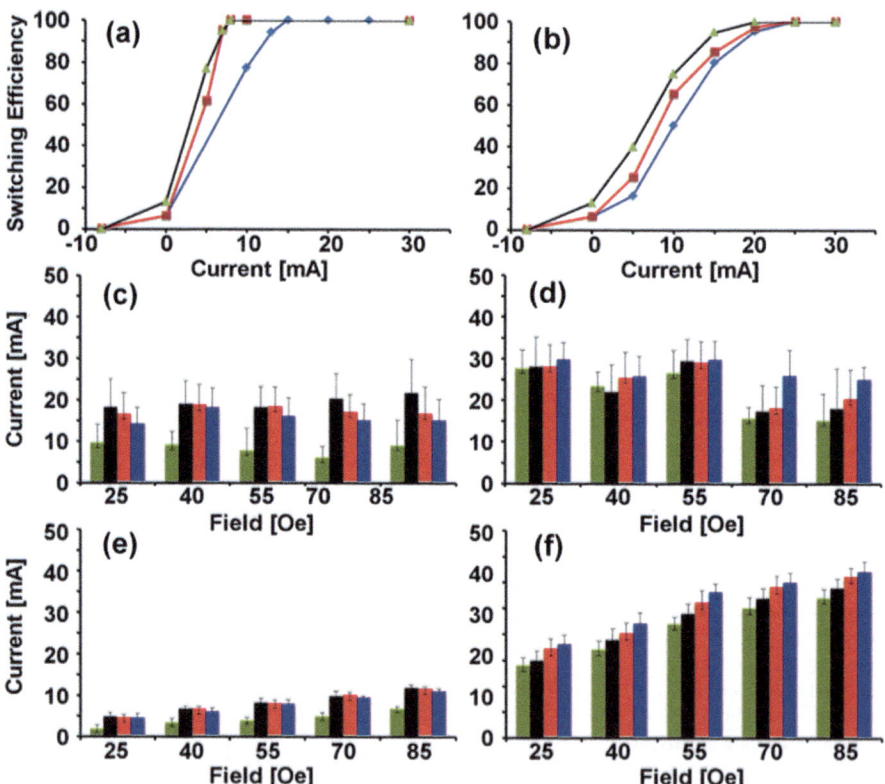

Fig. 5.15 Switching thresholds for transistors composed of whole disk tracks with a semiconducting gap for magnetic beads are presented. The experimentally obtained switching efficiencies in the dynamic transistor tests are illustrated for the **a** repulsive mode and **b** attractive mode at magnetic field frequencies of 0.2 Hz (blue), 0.5 Hz (red), and 0.8 Hz (black). The magnetic field magnitude is $H_{ext} = 45$ Oe. The experimentally obtained switching thresholds in the static transistor tests for the **c** repulsive mode and **d** attractive mode are presented for device geometries of d = 8 μm and L = 6.5 μm (green bars), d = 10 μm and L = 7 μm (black bars), d = 10 μm and L = 8 μm (red bars), and d = 10 μm and L = 9 μm (blue bars). Theoretical results shown in **e** and **f** correspond to the modes of (c) and (d), respectively. The figure is taken from [7] with permission

number of successful transporting magnetic beads and magnetically labeled cells across the gap has been monitored. The resulting plots as a function of the applied electrical gate currents in an externally applied 45 Oe rotating magnetic field are shown in Fig. 5.14a and b. Movies of the bead and cell movements in the dynamic transistor tests can be seen at http://onlinelibrary.wiley.com/store/10.1002/adma. 201502352/asset/supinfo/adma201502352-sup-0004-S4.mov?v=1&s=91bb4a61c 15172a83d3d7331a851c20d6337c09a and http://onlinelibrary.wiley.com/store/10. 1002/adma.201502352/asset/supinfo/adma201502352-sup-0005-S5.mov?v=1&s= 43506fa18097f49012fce5a3651dba7f33cb1c37 from [7].

In the static test, the particles are placed at the gap while the external magnetic field is exposed to the chip parallel to the track axis (x-direction). Then, by slowly increasing the gate current, the minimum required currents for the particles and cells to cross the gap are identified. Movies for magnetic beads and magnetically labeled cells can be seen at http://onlinelibrary.wiley.com/store/10.1002/adma.201502352/ asset/supinfo/adma201502352-sup-0002-S2.mov?v=1&s=f2bff6b038cba347c1efc 9ef6138282d8492b290 and http://onlinelibrary.wiley.com/store/10.1002/adma.201 502352/asset/supinfo/adma201502352-sup-0003-S3.mov?v=1&s=39e08e9cab44 f0dbca5381d6ca181df61d3177de, respectively, from [7]. The obtained average switching current thresholds for both attractive and repulsive transistor modes for magnetized human CD4+ T cells are close to the ones of magnetic particles. The plots are not provided here but can be accessed in [7].

The FEM-based theoretical model can be used to compute the expected gate current switching thresholds. In stochastic studies, it is assumed that the particle diameter is a random variable, with a mean and standard deviation of $5.72 \pm 0.86\,\mu$m, which is derived by measuring it for 30 magnetic beads under a microscope. The results are provided in Fig. 5.15e and f. In the bar plots of Fig. 5.15, the black, red, and blue bars correspond to transistors with L, ranging from 7 to 9 μm. The gap size is fixed at 10 μm. The green bars in Fig. 5.15 correspond to transistors with a gap size of 8 μm and a gate displaced from the gap center by 6.5 μm. In all cases, the gate width is 3 μm.

The simulation results and experimental data show the same order of magnitude of the required switching currents. Based on these results, this required current is in the range of 10–30 mA. Also, simulations and experiments confirmed that small gap sizes need lower electrical current thresholds, the trend of which is observed in the repulsive transistor modes of Fig. 5.15c and e. The same trend is seen for the attractive transistor modes and the experiments with living cells (see [7] for plots).

In comparing the data for the static and dynamic tests, it becomes apparent that at low frequencies, the switching thresholds are commensurate. Thus, the simpler static tests are good enough for quantifying the transistor performance in this condition. But this analysis is not suitable for analyzing the transistor operation at higher frequencies. In this condition, dynamic transistor tests are needed.

At high frequencies, the switching thresholds drop, indicating that frequency can be considered as an additional control parameter to lower the required gate currents. A transistor in the repulsive mode with a gate current of ~8 mA operating at 0.5 Hz can switch magnetic beads with ~100% efficiency (Fig. 5.15a). But this transistor needs a gate current of ~15 mA to achieve complete switching at the driving frequency of 0.2 Hz. This result is consistent with one obtained from the static transistor test. Complete switching of cells is reported to be achieved at gate currents of ~20–25 mA for the 0.2 and 0.5 Hz driving frequencies, which is also consistent with the static transistor tests.

The opportunity of having a single transistor operating at both attractive and repulsive modes is fundamentally important and makes this transistor a good candidate for being used in arrays resembling memories in computers. This memory chip is composed of numerous addressable storage sites (i.e., capacitors) to store single particles and cells, as opposed to the electronic data in the computer memories. This chip works based on the full write/read operation (i.e., writing a bead or cell to the array and extracting it from the array) provided by dual-mode transistors. This memory, called a magnetophoretic random access memory, is introduced in the following chapters.

5.5 Bends

In magnetophoretic circuits, similar to electrical circuits, bend conductors are needed. In magnetophoretic circuits, offsets in conducting paths have shown to be challenging [8]. In the junction formed of two perpendicularly joined conductors, particles may be trapped and have difficulty being transported.

Consider a right-angle bend conductor with magnetic disks with a radius of 10 μm in a 60 Oe rotating magnetic field, as illustrated in Fig. 5.16. In Fig. 5.16a–d, the energy distributions for magnetic field angles with 45° steps are depicted. The desired particle trajectory is shown by the gray arrow in Fig. 5.16a. But the particle trajectory obtained from the simulations (following the energy well in blue) at the corner, which is depicted by the black dotted line, shows that the desired trajectory along the magnetic tracks is not achieved. Instead, the particle is trapped in the corner. This problem is due to the overlap of the energy wells at the corner, which gives the particle the chance of switching between the two magnetic disks at the corner (i.e., disks numbered 1 and 3 in Fig. 5.16b). The experimental particle trajectories agree with the simulation results. The corresponding experimental results for the particle trajectory predictions in Fig. 5.16a–d are shown in Fig. 5.16e–h, respectively.

To study this problem, the effect of particle and magnetic disk sizes on the energy distribution is investigated. These studies indicate that the problematic overlapping of the energy wells at the corner depends on the distance between the disks numbered 1 and 3 in Fig. 5.17 (i.e., pq in Fig. 5.17a). It also depends on the size of the particle. Hence, a dimensionless parameter $\alpha = pq/r_p$ is defined, where pq and r_p are the length of that line and the particle radius, respectively. In Fig. 5.17a, in which $\alpha < \alpha_{critical}$, the energy wells around disks 1 and 3 overlap, so the particle can switch between them. But in Fig. 5.17b, in which $\alpha > \alpha_{critical}$, the mentioned wells are distinct, and thus they do not hand over the particle between themselves. For the geometry, we have $\alpha_{critical}$ which is calculated to be approximately 3. So, the bend conductor illustrated

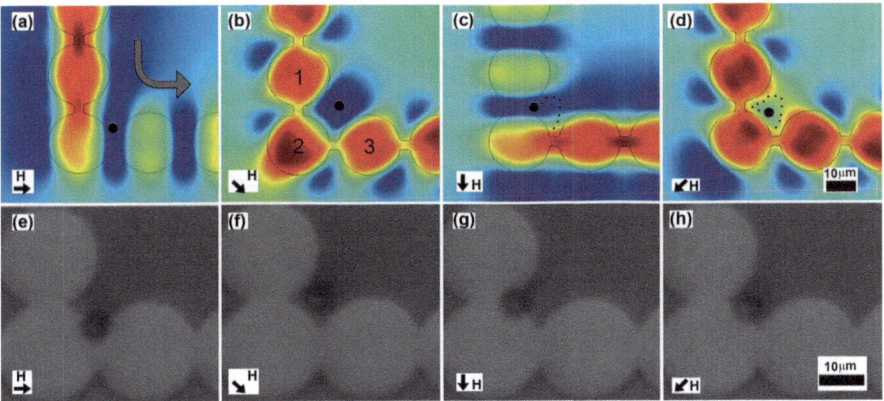

Fig. 5.16 A magnetophoretic right-angle bend conductor is shown. **a–d** The energy distributions and **e–h** the corresponding experimental results (top view) are shown. The field angles in each panel are depicted by the black arrows. In (a–d), the black circle and the black dotted lines depict the position and the trajectory of a particle, respectively. The gray arrow in (a) depicts the expected particle trajectory, which is not achieved. The blue and red areas depict the regions with low and high magnetic energies, respectively. The figure is taken from [8] under a Creative Commons Attribution (CC BY) license (http://creativecommons.org/licenses/by/4.0/)

in Fig. 5.17 appropriately transports the small particles; however, it has difficulties in moving the larger particles along the desired path. The curves plotted in Fig. 5.17c show the magnetic energy along the dotted line pq. These plots easily demonstrate the position of the energy wells with respect to each other at various heights. For instance, when $\alpha = 2.5$ or $\alpha = 4.17$, the two energy wells are overlapped or distanced, respectively. See this movie from [8] for an example of a particle having difficulties in movement on the corner: https://aip.scitation.org/doi/suppl/10.1063/1.5114883/ suppl_file/suppmovie1.wmv.

Close to the substrate, where the center of the small particles is located, narrow and deep energy wells form. For example, see the solid black line in Fig. 5.17c. But at higher heights, where the center of large particles is located, these energy wells get wider and shallower. For instance, see the dashed red line in Fig. 5.17c and compare it with the solid black line in this figure. This is the physical reason behind the effect of the particle size in its trajectory. For the large particles, the energy wells overlap, and their movement becomes challenging.

Large disks in designs enlarge the pq line and lead to a better α. Figure 5.17d illustrates the comparison of the pq length for a bend with small disks with the one with large disks. Table 5.1 illustrates the maximum particle radius for the desired particle transport with bends with disk radii of 5, 10, 15, and 20 μm. The strip length in between the disks is 3 μm.

So, we moved the energy wells far from each other by enlarging the magnetic disks. Another method to prevent the energy well overlapping challenge mentioned above is to use a magnetic disk at the corner larger than the other disks. Comparing p_1q_1 length with p_2q_2 length in Fig. 5.18a shows that in this design compared to the

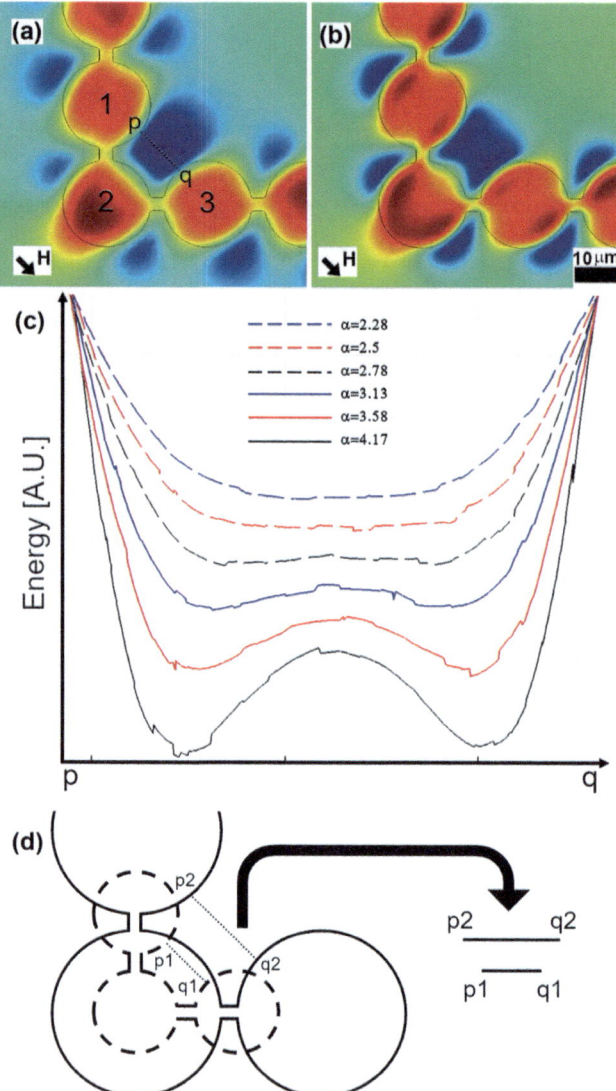

Fig. 5.17 The energy simulation results for a right-angle bend are presented. The energy distributions for **a** $\alpha = 2.5$ and **b** $\alpha = 4.17$ are shown. The external magnetic field directions are depicted with the little black arrows. The blue and red regions represent the area with low and high magnetic energies, respectively. **c** The magnetic energy along the pq line (in [a]) for α of 4.17 (solid black line), 3.58 (solid red line), 3.13 (solid blue line), 2.78 (dashed black line), 2.5 (dashed red line), and 2.28 (dashed blue line) are illustrated. **d** The distance between the two disks in two designs with small (i.e., p_1q_1) and large disks (i.e., p_2q_2) are compared. For easier comparison, these lines are redrawn on the right. The figure is taken from [8] under a Creative Commons Attribution (CC BY) license (http://creativecommons.org/licenses/by/4.0/)

Table 5.1 Approximate maximum particle radii for an appropriate particle transport based on the disk radii are shown. The connecting magnetic strips are 3 μm long

Corner magnetic disk radius (μm)	Maximum particle radius (μm)
5	2
10	4
15	6
20	7

The table is taken from [8] under a Creative Commons Attribution (CC BY) license (http://creativecommons.org/licenses/by/4.0/)

previous one, a better α is achieved. The energy distribution for this bend design is shown in Fig. 5.18b. Clearly, the energy wells are distinct. Figure 5.18c presents successful experimental particle transport using this bend. In Fig. 5.18d, energy along the p_1q_1 line for corner disks with various radii and particles of different sizes is plotted. Based on these results, $\alpha_{critical}$ is roughly equal to 2.

In Table 5.2, the maximum possible particle sizes for appropriate particle transport using bends with corner disks of various sizes are shown approximately. The designs with larger corner disks at the corner can move particles with a wider size range. See the movie from [8] for energy distribution in an external rotating magnetic field: https://aip.scitation.org/doi/suppl/10.1063/1.5114883/suppl_file/suppmovie2.wmv.

Another approach to overcome the particle trapping challenge in corners is to widen the bend angle (i.e., design an obtuse bend) [8]. Figure 5.19a compares a 90° bend with a 135° bend, drawn by the solid black and dashed black lines, respectively. Two important phenomena help this design to transport the particles well. First, in this design, the pq line is enlarged (compare p_1q length with p_2q length in Fig. 5.19a), which as discussed before prevents energy well overlaps. Second, the energy barrier provided by the magnetic disk at the corner (magnetic disk numbered 2 in Fig. 5.19a) prevents the particle from switching between the two energy wells of the magnetic disks numbered 1b and 3. The simulation results for the two bends of Fig. 5.19 are shown in Fig. 5.19c, d. The energy barrier of magnetic disk 2 is overlaid in Fig. 5.19a, with a zoomed view in Fig. 5.19b. It is obvious in Fig. 5.19a that the p_1q line crosses over the red region, which stands for the mentioned energy barrier. But the p_2q line crosses over the yellow area, which is a weak barrier and can be overcome with the energy wells, generated with disks 1b and 3 in the obtuse bend, a strong energy barrier (i.e., the red region in Fig. 5.19a) is produced by disk numbered 2, which prevents unwanted switching of the trapped particle between the two wells. But, the weak energy barrier in line p2q (i.e., the yellow region in Fig. 5.19a) can easily be overcome by the energy wells created by disks 1a and 3. Hence, the two energy wells may merge and switch the particle between themselves.

Energy distributions for multiple bend angles are presented in Fig. 5.19c–f. It is obvious that widening the bend angle leads to more distinct energy wells. The energy simulations have been repeated for particles of various sizes, the results of which are tabulated in Table 5.3. Based on these results, a bend of 135° transports particles up to 14 μm in radius, which covers a wide range of particles of interest,

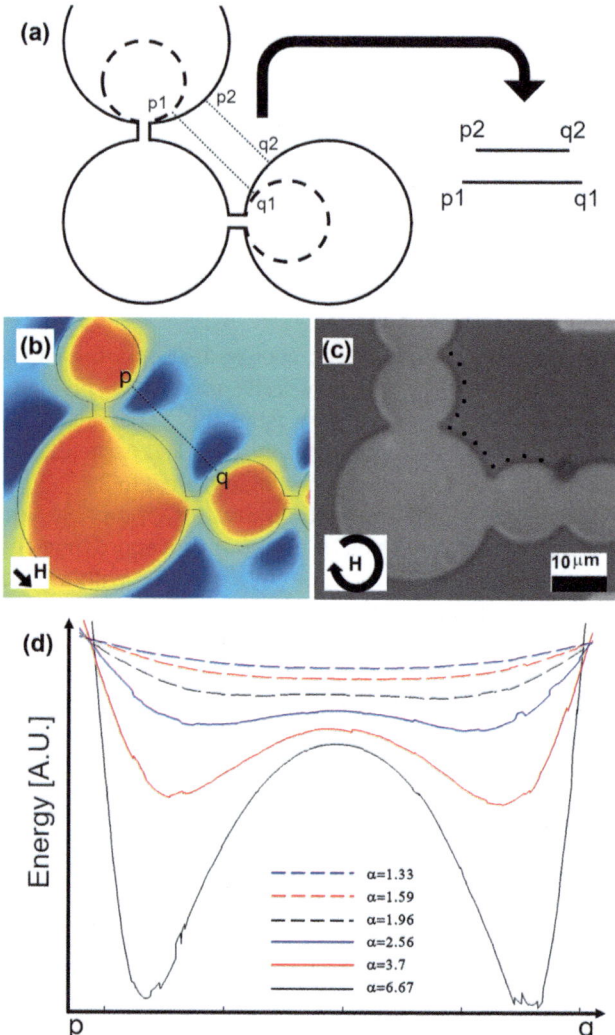

Fig. 5.18 The bend design based on a large disk at the corner is shown. **a** The distance between the disks in a bend with a large corner disk (i.e., p1q1) is larger than the one in a design with magnetic disks of similar size (i.e., p2q2). For easier comparison, these two lines are drawn on the right. **b** The energy distribution for this design is illustrated. The black arrow depicts the magnetic field direction. The blue and red areas stand for the regions with low and high energies, respectively. **c** The experimental particle trajectory is shown with the black dotted line. The rotation of the external magnetic field is depicted with the black circular arrow. **d** The magnetic energy along the dotted line pq (in (b)) for α of 6.67 (solid black line), 3.7 (solid red line), 2.56 (solid blue line), 1.96 (dashed black line), 1.59 (dashed red line), and 1.33 (dashed blue line) are plotted. The figure is taken from [8] under a Creative Commons Attribution (CC BY) license (http://creativecommons.org/licenses/by/4.0/)

Table 5.2 The approximate maximum possible particle size for an appropriate particle transport for bends with various corner disk radii are listed. The length of the strips connecting the magnetic disks and the radius of magnetic disks other than the corner disk are 3 μm and 10 μm, respectively

Corner disk radius (μm)	Maximum particle radius (μm)
10	4
15	8.5
20	11

such as most cells. Two or more obtuse bends can be combined to form a miter bend or long-radius elbow. Figure 5.19g illustrates the chip area required for three different bend angles, indicating that wider bends need more surface area on the chip. Thus, a tradeoff exists between the two parameters of the transportable particle size range and the design compactness. The 135° bend is an optimum candidate offering a good balance between the two parameters. The simulation results and experimental microscopy images for a 135° bend are shown in Fig. 5.20. This figure shows that the bend transports a single cell smoothly.

The particle transports on multiple bend designs and their corresponding transport efficiencies are illustrated in Fig. 5.21(a–d) and Fig. 5.21(e–h), respectively. In these experiments, magnetic beads of different sizes of 5–5.9 μm (named 5 μm in Fig. 5.21), 10–13.9 μm (named 10 μm in Fig. 5.21), and 14–17.9 μm (named 15 μm in Fig. 5.21) and magnetically labeled CD4+ T cells are used. The effect of the applied frequency is also studied. The design in Fig. 5.21a shows, as discussed before, that a large disk at the corner helps transport particles of different sizes. The design in Fig. 5.21b shows a bend composed of six 165° bends, which transports various particles smoothly. This design shows a better particle transport at higher frequencies (compare the gray bars in Fig. 5.21e, f). Similar results are also seen in the design shown in Fig. 5.21c which is composed of two 135° bends. The unoptimized original bend is illustrated in Fig. 5.21d. It shows the lowest particle transport efficiencies, as plotted in Fig. 5.21h. See movies from [8] for the operation of a few offsets based on the obtuse bends: https://aip.scitation.org/doi/suppl/10.1063/1.511 4883/suppl_file/suppmovie3.wmv; https://aip.scitation.org/doi/suppl/10.1063/1. 5114883/suppl_file/suppmovie4.wmv; https://aip.scitation.org/doi/suppl/10.1063/ 1.5114883/suppl_file/suppmovie5.wmv

In a fixed driving frequency, the particles need to travel a longer perimeter around the large magnetic disks compared to the smaller ones, at a certain time. In other words, they need to move faster. At higher velocities, the viscous drag force on particles and their phase lag with respect to the external magnetic field increases. Thus, in order to prevent shifting from the phase-locked regime into the phase-slipping regime, where the particles cannot be synced to the external magnetic field, the operating frequency must be kept at low values. Also, $F_{drag} = 6\pi \vec{v} \eta_f r_p$ mentions that the drag force for large particles is higher. Hence, the bends with large disks can

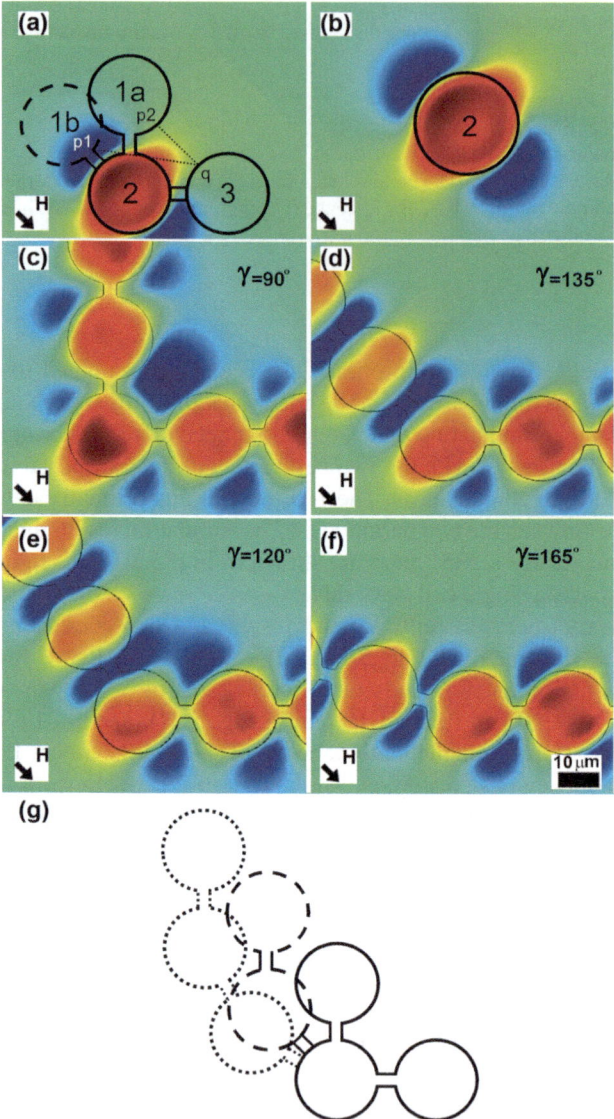

Fig. 5.19 The obtuse bend design is illustrated. **a** A 135° bend design, composed of disks 1b, 2, and 3, is overlapped on a 90° bend, composed of disks 1a, 2, and 3 for comparison. The energy simulation of disk 2 is overlaid. The p_1q and p_2q lines connecting disk 3 to disks 1b and 1a, respectively, are shown. **b** The energy landscape for disk 2 is shown. **c** The energy distribution for a 90° bend is shown. **d** The energy distribution for a 135° bend is illustrated. **e** The energy distribution for a 120° bend is shown. **f** The energy distribution for a 165° bend is plotted. γ is the bend angle. The black arrow in each panel depicts the magnetic field direction. The blue and red regions stand for the areas with low and high energies. **g** Multiple bends designs are plotted to visualize and compare their occupying surface area on the chip. The figure is taken from [8] under a Creative Commons Attribution (CC BY) license (http://creativecommons.org/licenses/by/4.0/)

Table 5.3 The approximate maximum possible particle size for an appropriate particle transport for bends with various angles is listed. The length of the strips connecting the magnetic disks and the radius of magnetic disks other than the corner disk are 3 μm and 10 μm, respectively

Bend angle (°)	Maximum particle radius (μm)
90	4
105	7
120	9
135–180	14

The table is taken from [8] under a Creative Commons Attribution (CC BY) license (http://creativecommons.org/licenses/by/4.0/)

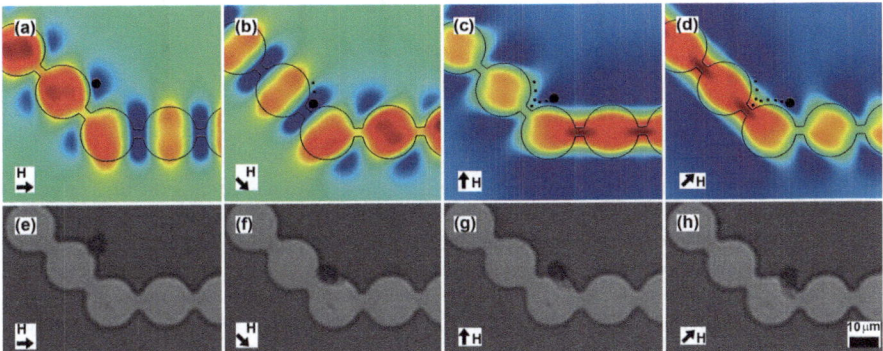

Fig. 5.20 A 135° Obtuse bend design is illustrated. **a–d** The energy landscapes for various magnetic field angles (depicted with the black arrows) are illustrated. The black circle and the dotted lines are the particle and its trajectory, respectively. **e–h** The corresponding experimental results for transporting a single living cell are presented (top view). The figure is taken from [8] under a Creative Commons Attribution (CC BY) license (http://creativecommons.org/licenses/by/4.0/)

only operate at low frequencies, and this is even more critical in transporting large particles.

Typically, the magnetic moment of the cells is lower than that of the magnetic beads. Hence, the cells experience a weaker magnetic force, and at high frequencies, they cannot follow the external magnetic field very well (see Fig. 5.21h).

Please note that the energy overlapping problem seen in bends is valid for the concave side of the bends. On the convex side of the bends, the energy wells are far away from each other. Hence, they can manipulate the particles smoothly. See this movie from [8] for an example: https://aip.scitation.org/doi/suppl/10.1063/1.511 4883/suppl_file/suppmovie4.wmv.

To conclude, similar to the electrical circuits, where a potential difference in conductors leads to electrical currents without requiring individual control signals for individual electrons, magnetophoretic circuits transport magnetic particles in parallel along the magnetophoretic conductors in an external rotating magnetic field. Bends are introduced to transport the particles at junctions without unwanted closed-loop particle trajectories. In addition to the magnetophoretic conductors, the diodes with

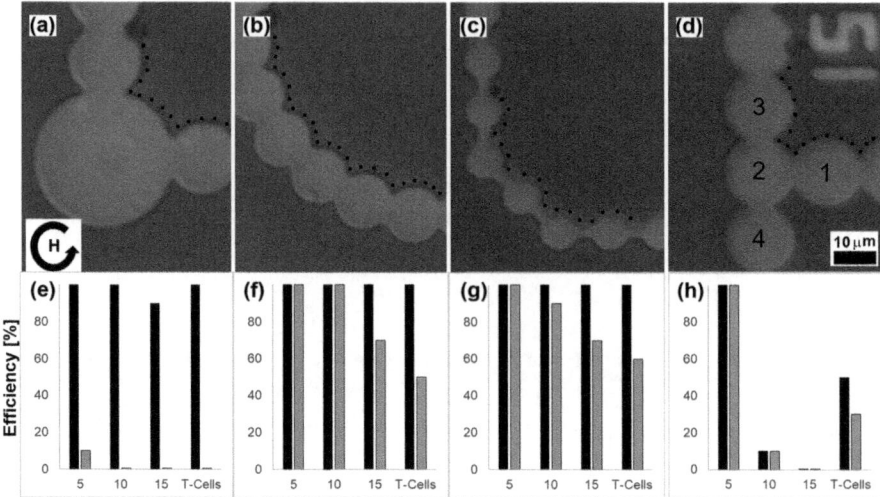

Fig. 5.21 Bend designs and their efficiencies are compared. **a** A bend equipped with a large disk at the corner, **b** a long radius elbow, **c** a miter bend, and **d** a 90° bend are presented. The particle trajectories and the rotation of the external field are represented with the black dotted lines and the black circular arrow, respectively. **e–h** The particle transport efficiencies for the 5 μm, 10 μm, and 15 μm magnetic particles as well as magnetically labeled human T cells at external magnetic field frequencies of 0.2 Hz (black bars) and 0.5 Hz (gray bars) are shown. See the text for the particle size distributions. The figure is taken from [8] under a Creative Commons Attribution (CC BY) license (http://creativecommons.org/licenses/by/4.0/)

the ability to transport the particles unidirectionally were investigated. The capacitors were also studied. It was shown that these circuit elements store the particles and can be used as single-cell compartments for further biological studies. These components are passive circuit elements in which no electrical signals are needed.

In addition to the passive circuit elements, active magnetophoretic circuit elements are introduced. Several types of magnetophoretic transistors capable of switching the trajectory of the magnetic particles are designed. The required gate currents for switching the particles and their efficiencies have been studied. The best transistors are the ones with the lowest required gate currents and with a tight spread of the particle trajectories. Working with low gate currents becomes important when in the integrated magnetophoretic circuits, the transistors are assembled in series, needing higher total voltages to sufficiently trigger their gates. Also, the tight spread of the trajectories shows the repeatability of the experiments and the reliability of the device. It is reported that the electrical gate current thresholds for the transistors working in the repulsive mode are typically lower than the required electrical currents for the transistors operating in the attractive mode.

References

1. Stratton, J. A. (1941). *Electromagnetic theory*. McGraw-Hill Book Company, Inc.
2. Abedini-Nassab, R., Ding, X., & Xie, H. (2022). A novel magnetophoretic-based device for magnetometry and separation of single magnetic particles and magnetized cells. *Lab on a Chip, 22*(4), 738–746.
3. Abedini-Nassab, R. (2022). Magnetometamaterials: Metamaterials with tunable magnetic matter conductivity. *Physical Review Applied, 17*(1), 014020.
4. Lim, B., et al. (2014). Magnetophoretic circuits for digital control of single particles and cells. *Nature Communications, 5*, 3846.
5. Abedini-Nassab, R., et al. (2014). Optimization of magnetic switches for single particle and cell transport. *Journal of Applied Physics, 115*(24), 244509.
6. Tejedor, M., et al. (1995). External fields created by uniformly magnetized ellipsoids and spheroids. *IEEE Transactions on Magnetics, 31*(1), 830–836.
7. Abedini-Nassab, R., et al. (2015). Characterizing the switching thresholds of magnetophoretic transistors. *Advanced Materials, 27*(40), 6176–6180.
8. Abedini-Nassab, R., & Shourabi, R. (2019). Bends in magnetophoretic conductors. *AIP Advances, 9*(12), 125121.

Chapter 6
Magnetophoretic Circuits Operating in a Tri-Axial Magnetic Field

In an in-plane magnetic field, the opposite poles of the magnetic particles, which can be considered dipoles, are aligned toward each other. Thus, when two particles come into contact attract each other and form a particle cluster (see Fig. 6.1a). Since the goal is to use the magnetophoretic circuits as a single-cell analysis tool, forming particle clusters is undesired. Furthermore, after a while, by approaching the other particles, the cluster may grow and alter the device performance.

To avoid this potential problem, the particles can be biased by introducing an additional vertical field to the system [1], as shown in Fig. 6.1b, to move the opposite poles away from each other. In a tri-axial magic field with a conical angle α, smaller than a critical angle (see Fig. 6.1c), the force between particles is repulsive. From experiments, this critical angle is found to be $\alpha = 54°$.

The circuit elements designed to operate in a 2D field do not work properly in a 3D field. Thus, a whole set of new circuit elements are required. But before investigating these circuit elements, let's discuss the magnetometamaterials composed of 2D arrays of magnetic disks, which, as mentioned before, can operate in a tri-axial magnetic field.

At low frequencies, it is obvious that the particles circulate a single disk. But increasing the magnetic field frequency leads to closed-loop particle trajectories between the adjacent magnetic disks both in x- and y-directions, every quarter of the period. Figure 6.2a, b show the waveforms of the magnetic field components and the resulting particle trajectory, respectively [2]. The frequency in this experiment is 0.3 Hz. The red dots in Fig. 6.2b show the real particle trajectories obtained from the experiments.

So, in a 2D magnetic field, particles experience closed-loop particle trajectories. This challenge cannot be answered even by increasing the frequency. This problem is due to the fact that magnetic particles periodically switch between the north and south poles of the adjacent magnetic disks without penalty [2]. The energy landscape simulation results for four sequential time points in Fig. 6.3a–d show this problem in a clearer way. In this figure, the red dots stand for the particle experimental trajectory which is overlaid on the simulated energy landscape. Thus, if we can somehow

Fig. 6.1 A schematic for magnetic particle alignment in a magnetic field is shown. **a** The particles are exposed to an in-plane 2D magnetic field. **b** The particles are exposed to a tri-axial magnetic field with a vertical bias component. N and S depict the north and south poles, respectively. Blue, red, and black arrows stand for the attractive force, repulsive force, and superimposed magnetic field direction, respectively. In **c** α is the conical angle of the magnetic field

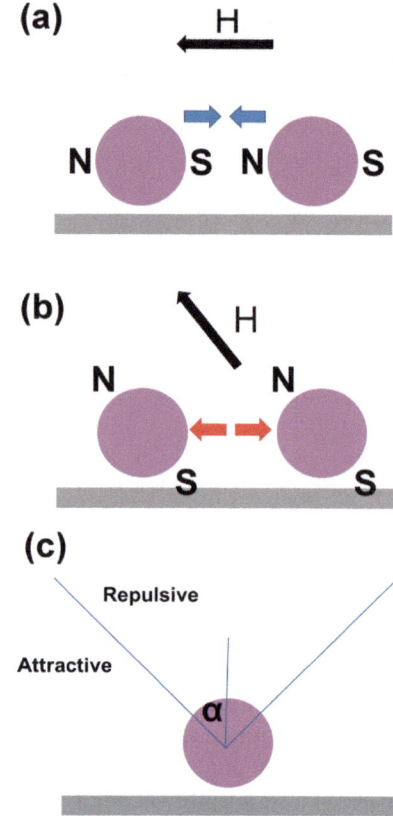

eliminate the energy well of specific magnets at specific time points, the problem can be solved. This goal can be achieved by adding a vertical magnetic field bias to the system, which is the same solution for preventing particle cluster formation as explained earlier.

The superimposed vertical magnetic field bias turns one of the energy wells around a magnetic disk into an energy peak. Thus, as shown in Fig. 6.3e–h, single energy wells and their following particles circulate single disks. But this is again a closed-loop particle trajectory. Open trajectories need tuned rectangular waveforms as the bias vertical magnetic field to appear/disappear the energy wells at the desired time points. For example, as seen in Fig. 6.3d, in a clockwise in-plane rotating field super-imposed with a square wave with the timing shown in Fig. 6.3c, the particles move in the positive x-direction. As another example, diagonal trajectories are achieved by applying an asymmetric rectangular waveform as the vertical bias field, as shown in Fig. 6.3e, f.

To systematically study the particle transport, again the dimensionless parameter β can be used. Based on the experimental results in [2], at frequencies in the range of $f_{cm} < f < f_{cs3}$, for $\beta \geq 0.35$ open particle trajectories (i.e., conducting regime) are seen.

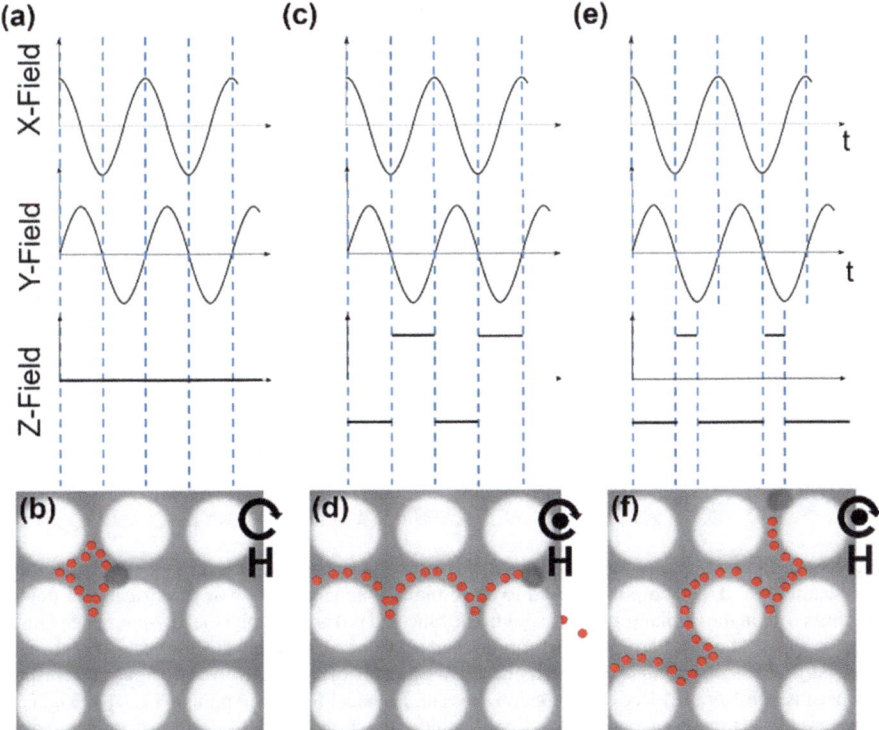

Fig. 6.2 The operation of two-dimensional magnetometamaterials is shown. **a, c, e** The magnetic field components are plotted. **b, d, f** The experimental particle trajectories are shown. The red dotted lines show the experimental particle trajectories. The circular arrow and the black circle depict the external magnetic field rotation and the vertical magnetic field bias, respectively. The figure is reprinted with permission from [2] Roozbeh Abedini-Nassab, Physical Review Applied, 17, 014020, 2022. Copyright (2022) by the American Physical Society. https://doi.org/10.1103/PhysRevApplied.17.014020

At too low frequencies (i.e., $f < f_{cm}$), the particle circulates a single disk, while at too high frequencies (i.e., $f > f_{cs3}$), the particle does not move (i.e., is trapped). For $\beta < 0.35$, the gap is too large for the particle to cross it and so close trajectories are seen. Thus, by choosing an appropriate β good conditions for the magnetometamaterials can be chosen.

As mentioned before, the movement of the particles is similar to the movement of the electrons in electric materials. The particle velocity versus the applied magnetic field frequency curves show that particle movement also obeys Ohm's law in electrical circuits. Figure 6.4 depicts the linear relationship between the magnetic particle current and the frequency. At high frequencies in which the particles enter the phase-slipping regime, Ohm's law is not followed anymore.

It was mentioned earlier that including a vertical bias field offers the advantage of inhibiting particle cluster formation. But in order to have particles transported along

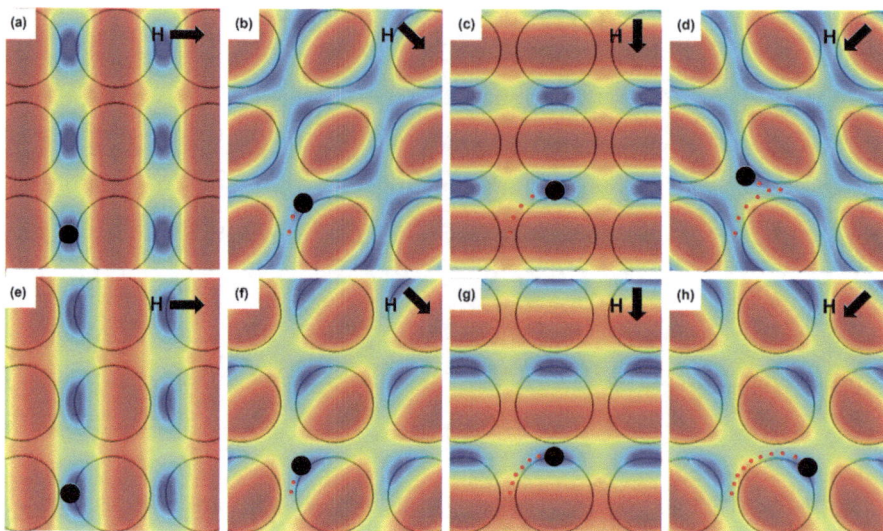

Fig. 6.3 Simulation results for the energy landscape of a two-dimensional magnetometamaterial are illustrated. **a–d** The chip is exposed to an in-plane magnetic field. **e–h** A vertical bias field is superimposed on the in-plane magnetic field. The blue and red areas depict the low-energy and high-energy regions. The black arrows, black circles, and red dotted lines depict the field direction in each panel, particle position, and overlaid experimental trajectories, respectively. The figure is reprinted with permission from [2] Roozbeh Abedini-Nassab, Physical Review Applied, 17, 014020, 2022. Copyright (2022) by the American Physical Society. https://doi.org/10.1103/PhysRevApplied.17. 014020

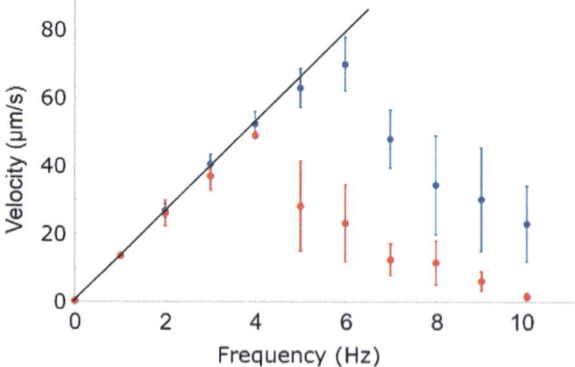

Fig. 6.4 Particle conduction as a function of frequency is shown. The blue and red dots show the data for particles with mean diameters of 2.8 and 5.7 μm, respectively, with the trajectory shown in Fig. 6.2d. The figure is reprinted with permission from [2] Roozbeh Abedini-Nassab, Physical Review Applied, 17, 014020, 2022. Copyright (2022) by the American Physical Society. https:// doi.org/10.1103/PhysRevApplied.17.014020

magnetometamaterials based on magnetic disks, as explained above, tuned magnetic fields are needed. Since providing these control signals in some labs may not be easy, other magnetic patterns are designed to be used as magnetophoretic conductors operating in a single in-plane rotating field superimposed with a constant vertical bias field.

In designing the magnetic patterns, if all the geometries have positive curvature (e.g., the magnetic disks), the energy wells circulate closed-loop trajectories. But if we include alternating sections of positive and negative pattern curvatures in the magnetic track geometry, open particle trajectories may be achieved.

Old logic operation and digital information storage systems were based on magnetic bubble technology, from which ideas for transporting magnetic particles are borrowed. In this technology, which was explored in the 1970s and 1980s [3–8], a vertical field creates magnetization domains in iron garnet films. These domains, called magnetic bubbles, represent data bits. The bubbles were moved along desired directions using magnetized magnetic tracks in a time-varying tri-axial magnetic field. A similar idea is more recently used to manipulate ferrofluid droplets immersed in oil [9].

This idea is also used in designing magnetophoretic circuits operating in a tri-axial magnetic field to manipulate magnetic particles. But there are some fundamental differences to be noted. Since the mechanical drag in the microfluidic environments is much higher, the operating frequencies in the magnetophoretic circuits are in the range of a Hertz (Hz), which is orders of magnitude lower than the operating frequencies in the magnetic bubble technology (i.e., kHz). Moreover, because the mobile components (i.e., the spherical particles in the magnetophoretic circuits and the cylindrical magnetic bubbles in the garnet films) have different geometries, the required magnetic fields for their transportation are different. Some other magnetic conductors, which were not used in the magnetic bubble technology, are also introduced, which show promising results.

6.1 Conductors

A sample design of a geometry having alternating sections of positive and negative pattern curvatures is shown in Fig. 6.5a. This design, called "drop-shape", enables magnetized particles and cells to move in open trajectories. Other geometries having alternating positive and negative curvatures also have the potential to transport magnetic particles. For example, the designs illustrated in Fig. 6.5b–f are adapted from magnetic bubble technology [3–8] to be used as a magnetophoretic conductor.

The numbers in Fig. 6.5 stand for the sequence of particle stable positions in a magnetic field with 90° intervals in the clockwise direction. The trajectories of several different particles on each conductor are plotted to better visualize their spread, where less spread indicates more reliable device operation. To see particle

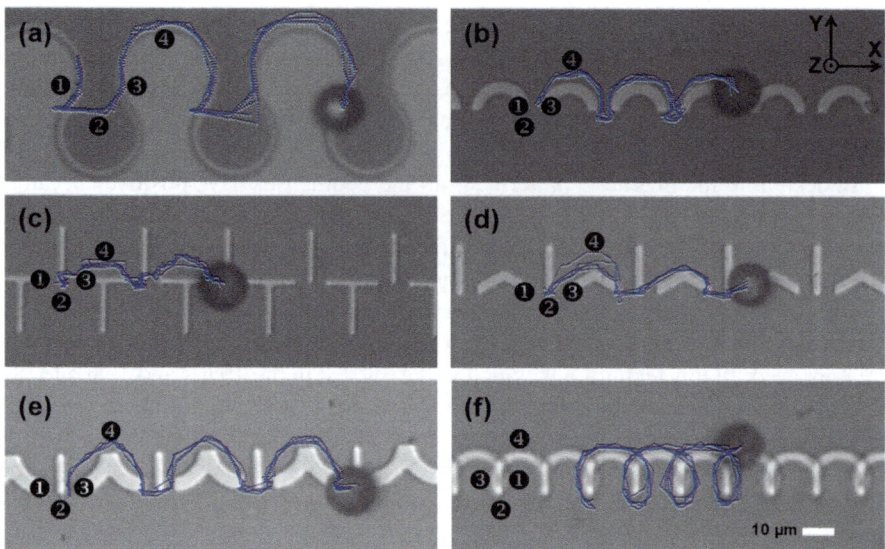

Fig. 6.5 Magnetophoretic conductors for transporting magnetic particles in a conical magnetic field are illustrated. Both horizontal and vertical magnetic field components are 45 Oe, and the driving frequency is 0.1 Hz. The numbers 1, 2, 3, and 4 stand for the sequence of the magnetic particle's stable positions in a clockwise rotating field. The 1 position corresponds to the case of + x directed in-plane magnetic field, and the other numbers correspond the sequential 90° clockwise rotations. The blue lines represent the experimental trajectories. The designs are called **a** drop-shape pattern, **b** C pattern, **c** TI pattern, **d** VI pattern, **e** YI pattern, and **f** spiral pattern. The figure is taken from [1] with permission

trajectory example movies see https://doi.org/10.1002/adfm.201503898/asset/sup info/adfm201503898-sup-0001-S1.wmv?v=1&s=92b43c36e6b950ef14649e396d6 6960d51c3763f and https://doi.org/10.1002/adfm.201503898/asset/supinfo/adfm20 1503898-sup-0002-S2.wmv?v=1&s=2059101dcd8c4b44a86d22b22768dd6e074c f90b.

The transport mechanism in a 3D (i.e., conical) magnetic field involves smooth transport along the sections of positive curvature followed by sudden transitions at sections of negative curvature. The previously defined parameter β can be used to systematically study the transport properties. To achieve open particle trajectories in most of the conductor geometries shown in Fig. 6.5b–f, $\beta \geq 2.5$ is needed [1]. But since the fabrication of small gaps using traditional lithography tools is not usually easy, this large ratio may become challenging. However, $\beta > 0.07$ is sufficient for the drop-shape pattern shown in Fig. 6.5a and makes it a good candidate to be further studied. Also, note that mirror symmetry is seen in all the designs shown in Fig. 6.5. Thus, in a magnetic field with the opposite direction, they can move the particles in the opposite direction. That means these are bidirectional conductors.

The conductivities of the different magnetic track designs shown in Fig. 6.5 are compared by measuring the maximum velocity of 15.6 μm magnetic beads on them.

Table 6.1 The maximum velocities obtained for particles with 15 μm diameter on various magnetophoretic conductors are listed

Type	Maximum speed ($\mu m\ s^{-1}$)
Drop-shape pattern	22.5
C pattern	7.8
TI pattern	9.6
VI pattern	6.0
YI pattern	3.5
Spiral pattern	3.0

The table is taken with permission from paper [1]

In each experiment, the frequency is increased, until the particle velocity drops and the maximum velocity is booked. The results are tabulated in Table 6.1. The drop-shape magnetophoretic conductor can conduct magnetic particles at the highest speeds, with a peak velocity of ~22.5 μm/s.

Since the drop-shape magnetophoretic conductor shows promising capabilities, it is explained here in detail. Figure 6.6 shows the magnetic particle trajectories for magnetic field cone angles in the range of $\alpha = 26°$–$90°$. At cone angles less than $30°$ (e.g., Fig. 6.6a for $\alpha = 26°$), magnetic particles with a mean diameter of 8.4 μm move in closed-loop trajectories around the head of the drop-shape. See the movie at https://doi.org/10.1002/adfm.201503898/asset/supinfo/adfm201503898-sup-0004-S4.wmv?v=1&s=09ccc7b55612f442d2096ee79be4afc56e4b2fe7 from [1]. In the cone angle in the range of $\alpha = 30°$–$60°$, the magnetic particles move in open trajectories (see Fig. 6.6b–d). For cone angles larger than $60°$ (Fig. 6.6e $\alpha = 63°$, and Fig. 6.6f $\alpha = 90°$), the magnetic particles move in closed-loop trajectories at the base of the drop. Also, see a movie at https://doi.org/10.1002/adfm.201503898/asset/supinfo/adfm201503898-sup-0005-S5.wmv?v=1&s=9c26de565eb9c4531d3b2eac6d9425a880b233f5 from [1]. The proper conditions for repeatable conduction are reported to be achieved at cone angles of $\alpha = 45°$–$50°$. As shown in Fig. 6.6d, at these cone angles, less spread of the particle trajectories is observed.

Figure 6.7 shows the FEM-based potential energy landscapes as a function of time, with a time step of $45°$, for an $\alpha = 45°$ field cone angle. The energy landscapes are plotted at the center of a particle with a diameter of 8.4 μm. If we follow the blue areas which depict the energy wells, we can predict the particle trajectories. Starting from Fig. 6.7a, the energy wells rotate around the sections of positive curvature at the head of the drop-shaped (see Fig. 6.7a–d). Then at a magnetic field angle of $\theta \approx 135°$ (see Fig. 6.7d), the track geometry curvature turns from positive to negative. Thus, the particle is repelled from the base of the conductor with negative curvature. In Fig. 6.7e, f, the energy wells and their following particles move horizontally to the next drop-shape magnet. And the same particle trajectory is achieved for the next period of the magnetic pattern. The blue dots in Fig. 6.7 show the experimental trajectories to compare with the simulation predictions. Also, see a related

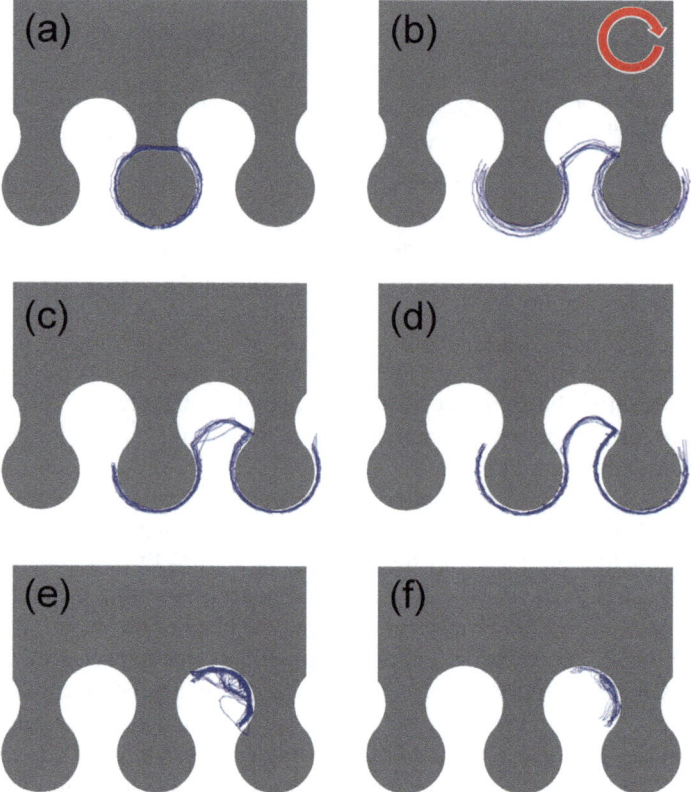

Fig. 6.6 The particle trajectories on drop-shape magnetophoretic conductors are shown. The trajectories are shown with blue lines. The particles are 8.4 μm in diameter. The magnetic field cone angles are **a** $\alpha = 26°$, **b** $\alpha = 37°$, **c** $\alpha = 45°$, **d** $\alpha = 53°$, **e** $\alpha = 63°$, and **f** $\alpha = 90°$. The red arrow stands for the field rotation sense. The figure is taken from [1] with permission

movie at https://doi.org/10.1002/adfm.201503898/asset/supinfo/adfm201503898-sup-0006-S6.wmv?v=1&s=08a1932ee67150341392504a00c171e264bec93e from [1].

Now let's see what happens if the applied magnetic field cone angle is small. Figure 6.8a–h illustrates the potential energy landscapes for a 10° field cone angle. The particle behaves similarly to what it does in a cone angle of $\alpha = 45°$ (see Fig. 6.8a–d). But, when the field angle is $\theta \approx 135°$, the strong vertical field leads to the creation of an energy well on the top of the magnetic pattern (see Fig. 6.8d–f). Thus, as opposed to moving along the magnetic track, the particles end up experiencing closed-loop circular trajectories around the head of the drop-shaped magnetic pattern. This behavior is also reported in the experimental studies (see blue dots in Fig. 6.8). Thus, this magnetic field cone angle is not appropriate for this magnetophoretic conductor. It is also reported that an in-plane rotating magnetic field does not lead to open trajectories of the particles [1].

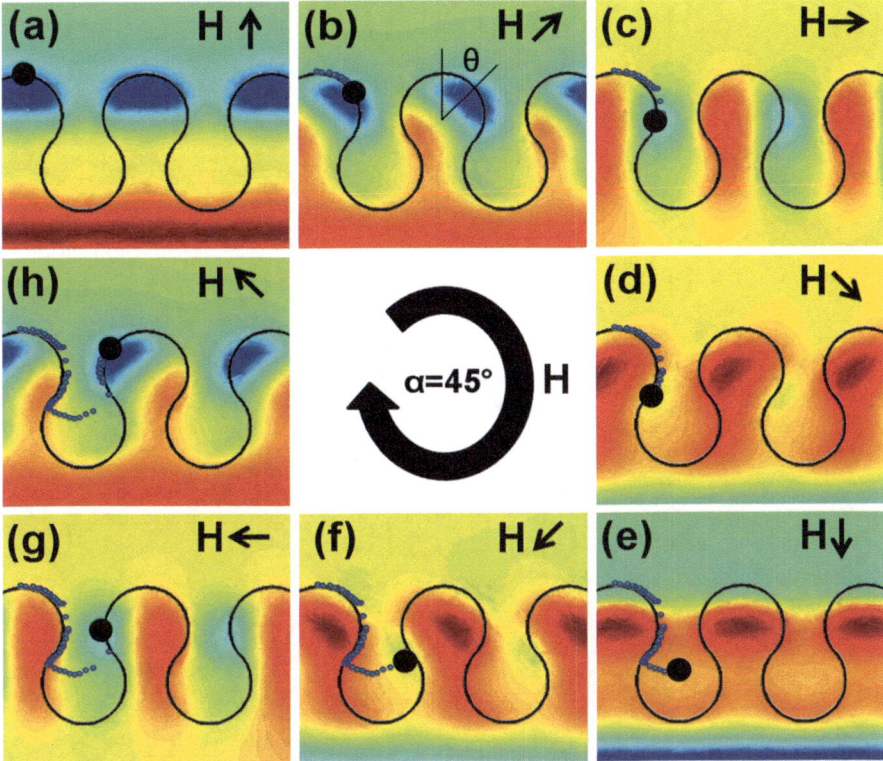

Fig. 6.7 The simulated energy landscapes for the drop-shape magnetophoretic conductor as a function of the in-plane magnetic field component, are plotted, where the cone angle is $\alpha = 45°$, the particle diameter is 8.4 µm, and the in-plane field angles are **a** $\theta = 0°$, **b** $\theta = 45°$, **c** $\theta = 90°$, **d** $\theta = 135°$, **e** $\theta = 180°$, **f** $\theta = 225°$, **g** $\theta = 270°$, and **h** $\theta = 315°$. The black arrows, the blue dots, the blue regions, and the red regions depict the external field direction, the experimental particle trajectories for 8.4 µm magnetic particles, the low energy regions, and the high energy regions, respectively. The figure is taken from [1] with permission

Thus, the magnetic field cone angle is an important parameter in the operation of drop-shaped magnetophoretic conductors. Another key parameter to be investigated here is the applied magnetic field frequency. Thus, the particle velocities as a function of the driving frequency for different cone angles are studied [1]. Figure 6.9 shows the results for various magnetic field angles, magnetic field strengths ranging from 25 to 100 Oe, and magnetic cone angles in the range of $\alpha = 26°-90°$. In these studies, the effect of particle diameter is also considered. The left column (Fig. 6.9a–c) shows the results for particles with a diameter of 5.6 µm, and the right column (Fig. 6.9d–f) depicts the results for particles with a diameter of 8.4 µm.

Based on these results, higher speeds are achieved at higher magnetic field strengths. Also, the larger particles can be manipulated at higher frequencies. The

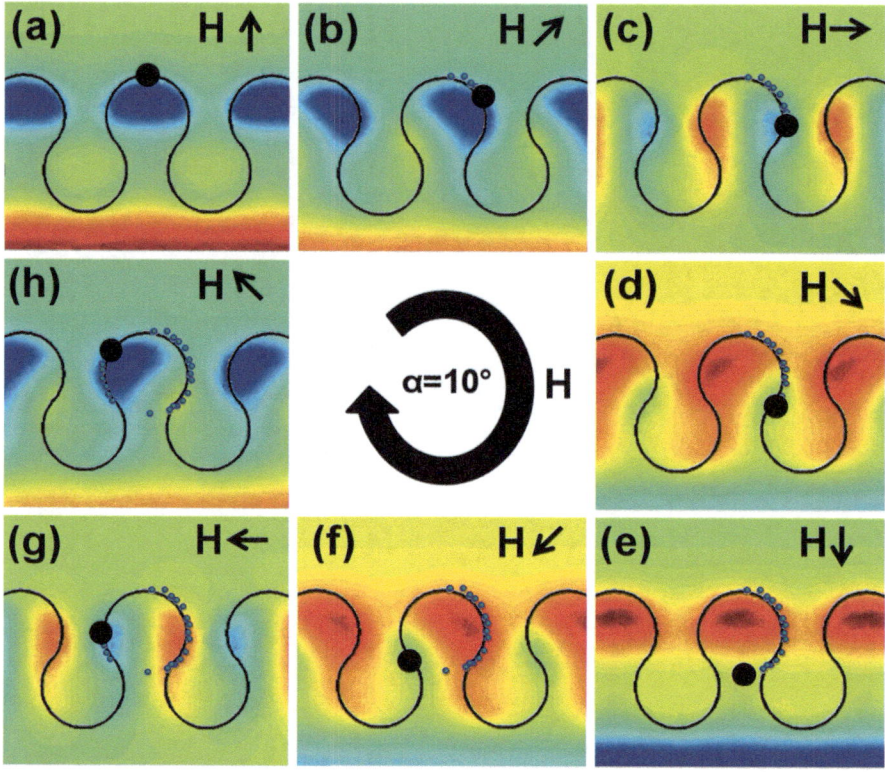

Fig. 6.8 The simulated energy landscapes for the drop-shape magnetophoretic conductor as a function of the in-plane magnetic field component, are plotted, where the cone angle is $\alpha = 10°$, the particle diameter is 8.4 µm, and the in-plane field angles are **a** $\theta = 0°$, **b** $\theta = 45°$, **c** $\theta = 90°$, **d** $\theta = 135°$, **e** $\theta = 180°$, **f** $\theta = 225°$, **g** $\theta = 270°$, and **h** $\theta = 315°$. The black arrows, the blue dots, the blue regions, and the red regions depict the external field direction, the experimental particle trajectories for 8.4 µm magnetic particles, the low energy regions, and the high energy regions, respectively. The figure is taken from [1] with permission

8.4 µm particles are transported at ~0.9 Hz; however, the 5.6 µm particles are transported at ~0.6 Hz. Particles could be transported in the cone angles of $30° < \alpha < 60°$.

Figure 6.10 shows the capability of the drop-shape magnetophoretic conductor in transporting magnetized cells. In this figure, the trajectories of magnetically labeled human T cells are illustrated with blue dotted lines. The close spread of the particle trajectories depicts the reliability of the drop-shape magnetophoretic conductor in transporting single cells. Also, you can see the movie showing the cell movement at https://doi.org/10.1002/adfm.201503898/asset/supinfo/adfm201503898-sup-0007-S7.wmv?v=1&s=041687ee712c22f57d3bee2bb720127ef0b0db53 from [1].

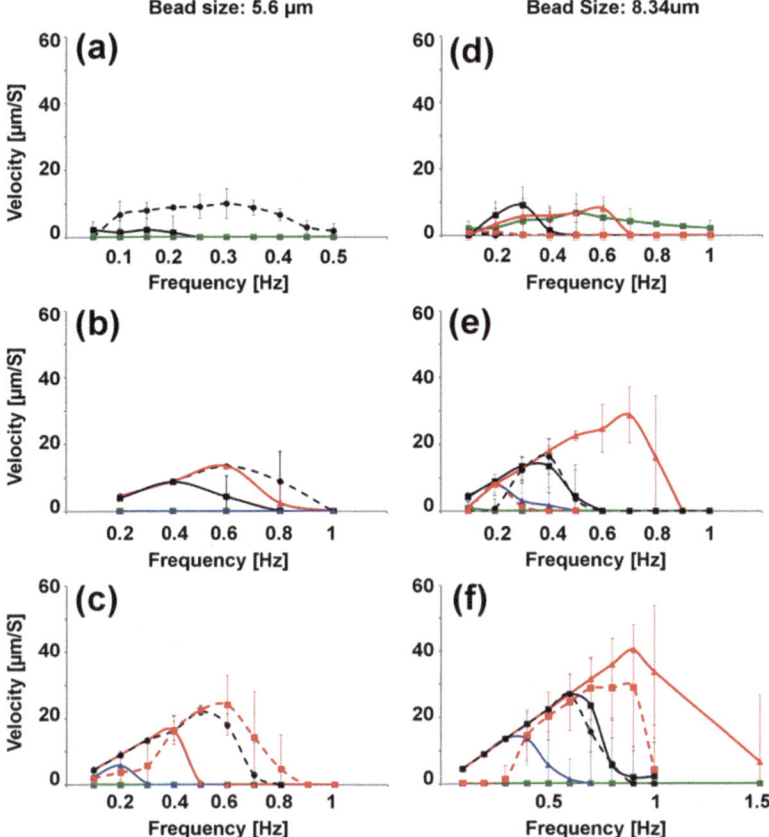

Fig. 6.9 The velocities of magnetic particles as a function of frequency for the drop-shaped magnetophoretic conductors are plotted. The left and right columns show the data for particles with diameters of 5.6 and 8.4 μm, respectively. The first, second, and third rows illustrate the data for 25 Oe, 50 Oe, and 100 Oe field strengths, respectively. The field cone angles are $\alpha = 26°$ (dashed red), $\alpha = 37°$ (dashed black), $\alpha = 45°$ (solid red), $\alpha = 53°$ (solid black), $\alpha = 63°$ (blue), and $\alpha = 90°$ (green). The figure is taken from [1] with permission

The second geometry to be studied here as a magnetophoretic conductor example is the one based on the TI bars. To explain the operation of this design easier, let's study the magnetic bar magnetization first.

In Fig. 6.11a, b, a magnetic bar in an in-plane magnetic field and after superimposing a vertical bias field, respectively, are shown [10]. The magnetic field distribution for the cases in Fig. 6.11a, b are shown in Fig. 6.11c, d, respectively. As expected, the addition of the vertical magnetic field eliminates one of the energy wells from one bar tip, leaving only a single tip with an energy well. Thus, considering a periodic magnetic pattern consisting of magnetic bars, in a rotating in-plane magnetic field, energy wells form at the bar tips sequentially. Now, if the bar tips are close to each other in the correct order, they can potentially transport the particles

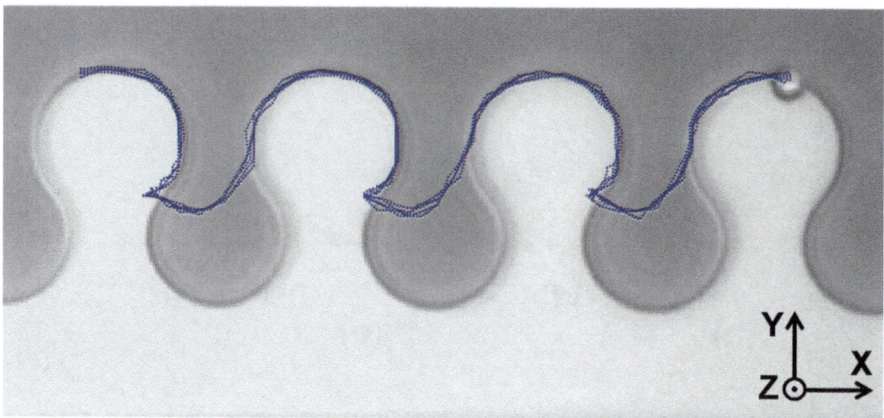

Fig. 6.10 The trajectory of magnetically labeled human T cells on a drop-shape magnetic conductor in a clockwise rotating magnetic field is shown. The in-plane and vertical field components of the magnetic field are Oe. The driving frequency is 0.1 Hz. The blue dotted lines stand for the particle trajectories. The figure is taken from [1] with permission

along the magnetic conductor track in a rotating field superimposed with a vertical bias field. Such a design is shown in Fig. 6.12. This figure shows the transport of the energy wells in this magnetophoretic conductor design at different time points. In these simulations, the cone angle $\alpha = 45°$, and the energy landscape is plotted at the center of a particle with a diameter of 8 μm. Starting from Fig. 6.12a, the magnetic particle is first on the bar tip aligned in the positive x-direction (see Fig. 6.12a), then it moves to the I bar (see Fig. 6.12c), and then the next T bar (see Fig. 6.12f). So, if we consider the particle position in different panels, we realize it moves from left to right as the magnetic field rotates clockwise. The particle trajectory is shown in Fig. 6.12 with the black dotted lines. Also, Fig. 6.12i shows the experimental particle trajectory along the proposed TI bar magnetophoretic conductor.

For this magnetophoretic conductor, the effect of the external magnetic field cone angle and frequencies are studied too. The efficiency of the TI magnetophoretic conductor is illustrated in Fig. 6.13. The plots are prepared for magnetic field cone angles ranging from $\alpha = 26°–63°$. Based on these results, the TI bar pattern works well at magnetic field cone angles of $\alpha = 30°–60°$. Also, low frequencies (i.e., less than 0.1 Hz) are needed to achieve good performances. Based on the reported experimental results, magnetic field strengths of ~80–100 Oe are suggested.

The TI bar conductor is also shown to manipulate live cells [10]. Figure 6.14 shows the movement of magnetically labeled AML cells on the T bar pattern, where the blue dotted lines are multiple cell trajectories. This figure shows that the cells move similarly to the magnetic particles without a problem.

The TI magnetophoretic conductors are also used for transporting droplets [9, 11]. Similar to what was mentioned before for the magnetic particles, the ferrofluid droplets follow the energy wells and move along the TI magnetic tracks (see Fig. 6.15a) [9]. The droplets have diameters of d = 300–1000 μm and are exposed to

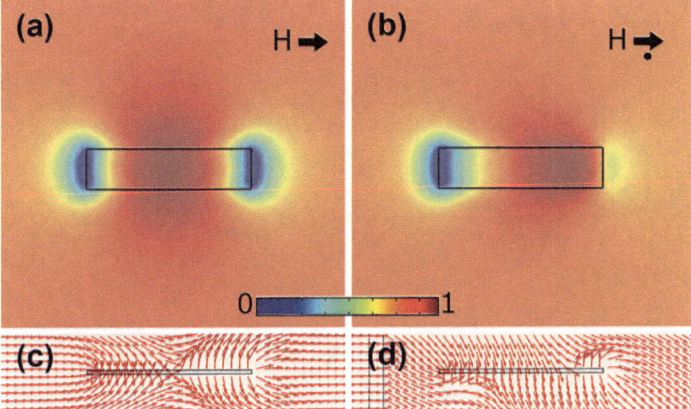

Fig. 6.11 The magnetic energy landscape for a magnetic bar is shown. **a** The bar is exposed to an in-plane magnetic field along its axis. **b** A vertical bias field is superimposed on the in-plane magnetic field. The blue and red areas show the regions with low and high magnetic energies, respectively. The black arrows and the black dot represent the magnetic field direction and the vertical bias field, respectively. Cross-sectional views of the magnetic field distribution for the cases of (**a**) and (**b**) are plotted in (**c**) and (**d**). The figure is reprinted from [10] with permission from the Royal Society of Chemistry

a vertical bias magnetic field of 100–400 G (see Fig. 6.15b). Similar to the particles, the droplets move from one pole of one bar to the other one (see Fig. 6.15c). The experimental setup is shown in Fig. 6.15f. Figure 6.15d shows a schematic of the droplet in the chip.

In all the systems discussed above, the moving objects (particles) are magnetic, and the surrounding media is non-magnetic. But water droplets with magnetic susceptibility of -10^{-5}, which can be considered non-magnetic with a good approximation, have also been manipulated in hydrocarbon oil-based ferrofluid [11]. Hence, in this system, the surrounding media is a paramagnetic material with initial susceptibility of ~3. In this platform, the water droplets, as discussed in Sect. 1.1.1.3 Droplet-based microfluidics, can perform as carriers.

6.2 Diodes

The magnetophoretic diode is the magnetophoretic circuit element that unidirectionally transports the magnetic particles and magnetized cells [1]. Two examples of these circuit elements are illustrated in Fig. 6.16. These elements are engineered by introducing an asymmetry to their design. In a clockwise rotating magnetic field (the left column in Fig. 6.16), the magnetic particles move in open trajectories (forward mode); however, in a counterclockwise rotating magnetic field (the right column in Fig. 6.16), the magnetic particles circulate in closed orbits (reverse mode). This

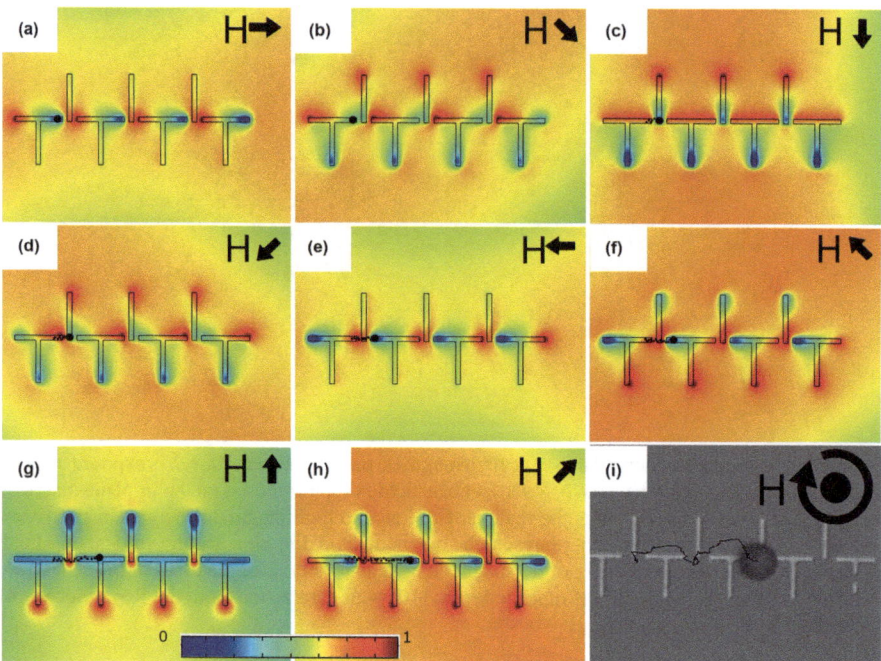

Fig. 6.12 **a–h** The simulation results for the magnetic energy landscape of the TI bar magnetophoretic conductor are plotted, for one cycle, at steps of 45°. The black arrow in each panel shows the direction of the in-plane magnetic field. A vertical magnetic field bias is also superimposed on the in-plane magnetic field. The black circle and the black dotted lines show the particle position and its trajectory at each step, respectively. The blue and red areas stand for the low energy and high energy regions, respectively. **i** The experimental trajectory of a single magnetic particle is shown. The figure is reprinted from [10] with permission from the Royal Society of Chemistry.

asymmetry in conduction is due to the deep energy well that forms close to the thick segment of the magnetic track.

When the magnetic field is in the positive x-direction, the particle is in position (1) in Fig. 6.16. After a magnetic field 90° clockwise rotation, the particle moves from the thin edge of the curved magnet to the adjacent I magnetic bar (i.e., position [2]). After the next 90° rotation of the magnetic field, the particle moves from the I bar to the next curved magnet. At this step, the magnetic field points to the negative x-direction, which is denoted as position (3). Then, the magnetic particle circulates the positive curvature of this magnet and arrives at position (4). Finally, by completing the cycle, the particle moves to position (1) on the adjacent unit cell (not numbered). Hence, in the forward mode, the particle experiences an open trajectory.

Now, in a field rotating counterclockwise (see the right column in Fig. 6.16), starting from the thinner segment of the curved magnet (position [1]), the particle circulates the positive curvature on top of the curved magnet. It arrives at position (2) and then position (3) after the field rotates 180°. At this point, since the deeper energy well exists in position (3), the particle does not cross over the gap to the

Fig. 6.13 The experimentally measured TI bar conductor efficiencies in transporting magnetic particles are plotted. The magnetic field strengths are **a** 25 Oe, **b** 50 Oe, and **c** 100 Oe. The mean particle diameters are 8.4 μm and the field cones are α = 26° (solid black), α = 37° (solid red), α = 45° (solid blue), α = 53° (dotted black), and α = 63° (dotted red). The figure is reprinted from [10] with permission from the Royal Society of Chemistry

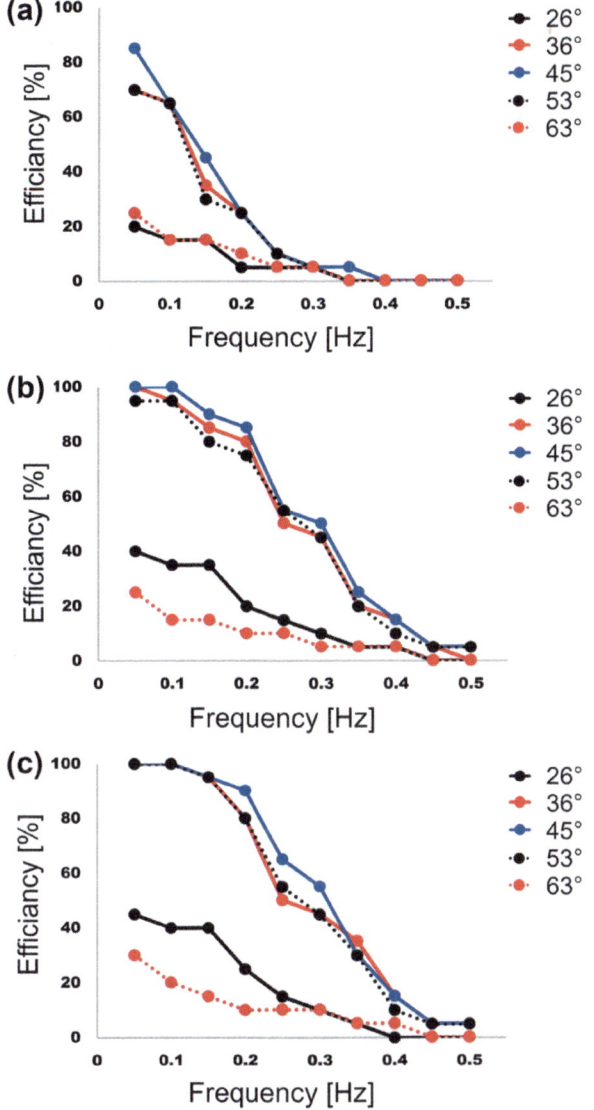

I-bar in position (4). Instead, the particle moves in a closed loop around the curved magnet. That means, in the reverse mode, the particle is not transported along the magnetophoretic diode. A movie of this behavior can be seen at https://doi.org/10.1002/adfm.201503898/asset/supinfo/adfm201503898-sup-0003-S3.wmv?v=1&s=2b19e0416f4c7a0186bbebc6be04aa4ba22ceb3e from [1].

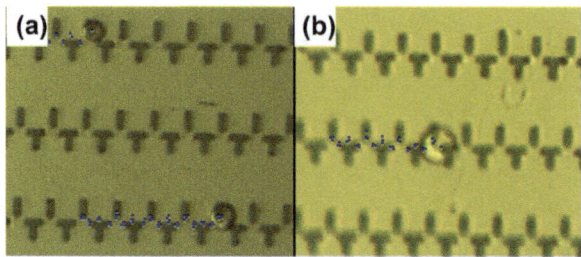

Fig. 6.14 a, b The trajectory (blue dotted line) of magnetically labeled AML cells on the TI bar magnetophoretic conductors in a clockwise rotating conical field with $\alpha = 45°$ is shown. The driving frequency is 0.05 Hz. The figure is reprinted from [11] with permission from the Royal Society of Chemistry.

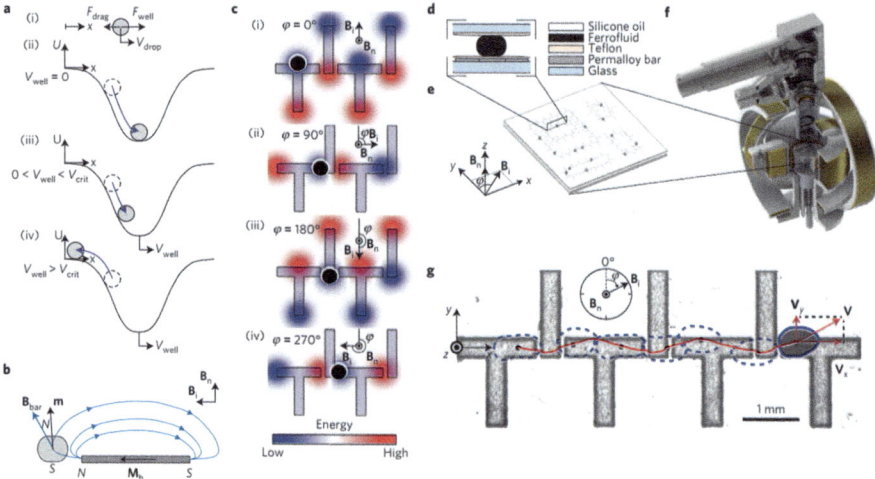

Fig. 6.15 Transport of droplets on TI magnetophoretic conductor is illustrated. **a** The droplets tend to move to the energy wells and follow them. **b** The schematic of the magnetic thin film and the droplet in a tri-axial magnetic field is shown. **c** The Energy distribution along the TI pattern is shown. **d** A schematic of a droplet inside a chip is illustrated. **e** A schematic of the chip is illustrated. **f** The experimental setup is depicted. **g** The trajectory of the droplet is depicted by the red line on the TI pattern. The figure is reprinted from [9] with permission. Copyright © 2015, Nature Publishing Group

6.3   Transistors

To switch the trajectory of the magnetic particles in a tri-axial magnetic field, a gate current is used to transfer them from one magnetic track to another one. This method is similar to the one in transistors operating in an in-plane magnetic field; however, here two drop-shape magnetic patterns are designed nearby [12]. Many gate geometries have shown to be promising results when used in these transistor

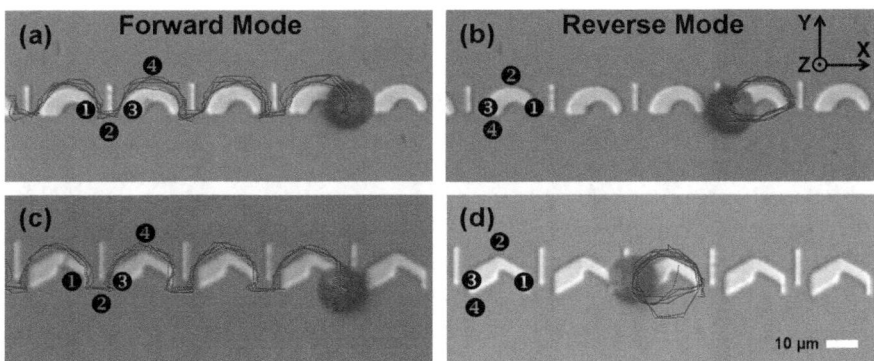

Fig. 6.16 Magnetophoretic diode examples are illustrated. **a, c** The diodes transport the particles in the forward mode when the field rotates clockwise. **b, d** But they do not transport the particles in the reverse mode when the field rotates counterclockwise. The horizontal and vertical field components are 45 Oe, and the frequency is set at 0.1 Hz. The blue lines show the experimental particle trajectories. The figure is taken from [1] with permission

designs. These magnetophoretic transistors are broadly classified into two general groups, which are discussed here.

6.3.1 Attractive Transistors

In the transistors called "attractive transistors", the gate forms a current loop on the magnetic pattern to momentarily attract the moving particle and prevent it from moving along its initial magnetic track by making it out of phase with the traveling magnetic energy along that track. Then by rotating the external magnetic field, when an energy well of the other magnetic track in proximity arrives at the junction, it gradually transfers the particle to that track, along which the particle continues its movement. Some representative experimental particle trajectories are shown with the blue dotted lines in Fig. 6.17. In these experiments, the magnetic field strength is 70 Oe with a cone angle of 45° and a driving frequency of 0.1 Hz. For switching the particles, electrical gate currents of 35–45 mA have been used.

6.3.2 Repulsive Transistors

The second category of transistors, called "repulsive transistors", have their gate wire parallel to the magnetic track direction (see Fig. 6.18). The current-carrying wire produces a repulsive magnetic force to repel the magnetic particle away from the initial magnetic track. This magnetic particle is then attracted by the second magnetic track and continues its movement along that track. In some designs, the

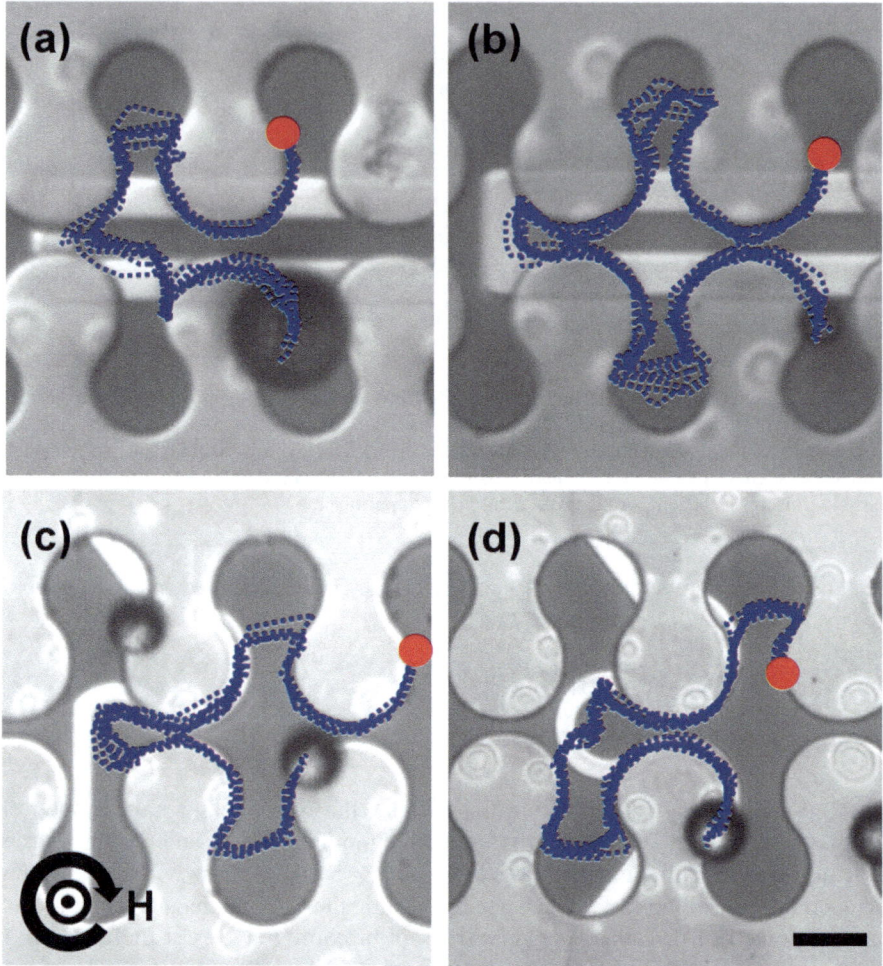

Fig. 6.17 Attractive transistors operating in a tri-axial magnetic field are illustrated. The magnetic field strength, cone angle, and frequency are 70 Oe, 45°, and 0.1 Hz, respectively. The required gate currents for switching the particles are reported to be **a** 35 mA, **b** 35 mA, **c** 45 mA, and **d** 40 mA, respectively. The dotted blue lines stand for the trajectories of magnetic particles. Each experiment is repeated multiple times. The red circles, the circular black arrow, and the dot in its center stand for the starting points of the particle trajectories, the rotation sense of the horizontal magnetic field component, and the vertical field, respectively. The figure is taken from [12] with permission from the Royal Society of Chemistry

wire is folded such that it creates an energy well on the sends track, helping it in attracting the particle (see Fig. 6.18a, b). Due to closer proximity to the repulsive section of the wire, the repulsion force is stronger than the attraction force on the other side.

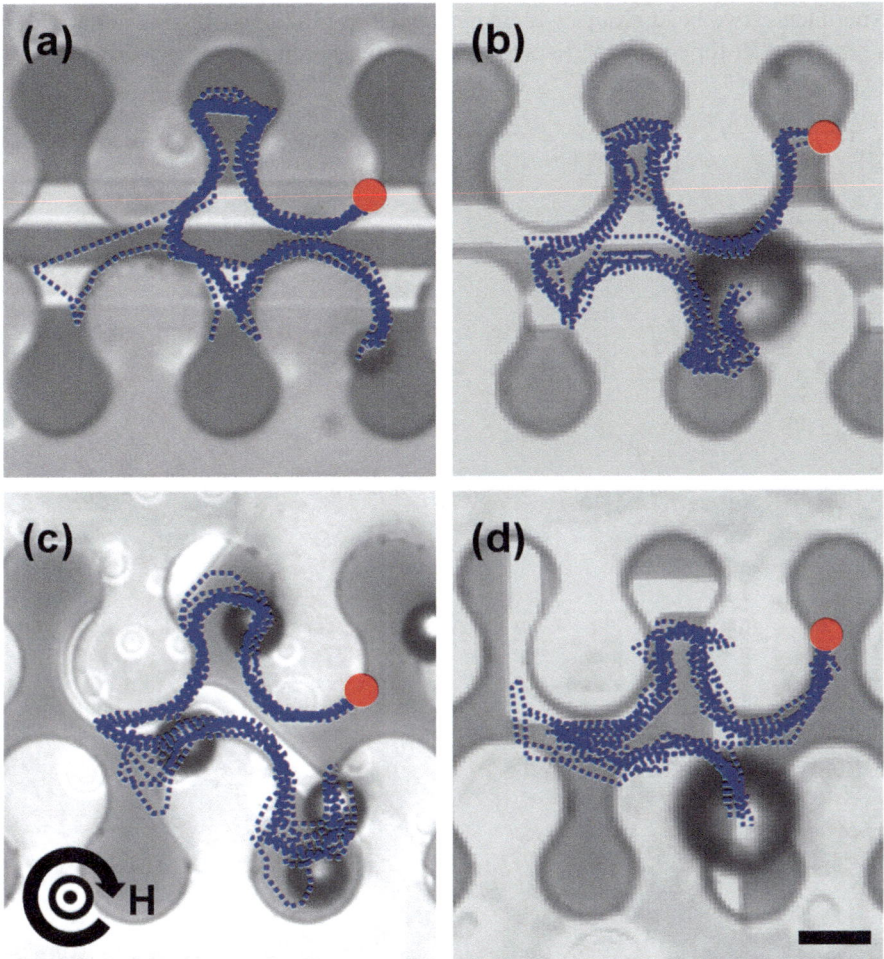

Fig. 6.18 Repulsive transistors operating in a tri-axial magnetic field are illustrated. The magnetic field strength, cone angle, and frequency are 70 Oe, 45°, and 0.1 Hz, respectively. The required gate currents for switching the particles are reported to be **a** 35 mA, **b** 30 mA, **c** 35 mA, and **d** 30 mA, respectively. The dotted blue lines stand for the trajectories of magnetic particles. Each experiment is repeated multiple times. The red circles, the circular black arrow, and the dot in its center stand for the starting points of the particle trajectories, the rotation sense of the horizontal magnetic field component, and the vertical field, respectively. The figure is taken from [12] with permission from the Royal Society of Chemistry

Figure 6.19 illustrates the quantification results for the mentioned transistors. In the reported analysis, the percentage of successful magnetic particle switching as a function of the applied electrical gate currents is investigated. The chosen magnetic field, cone angle, and frequency are 70 Oe, 45°, and 0.1 Hz, respectively. In these experiments, two bead groups with mean diameters of 8.4 and 15.6 μm are used. It is reported that for each experiment at least ten measurements are performed.

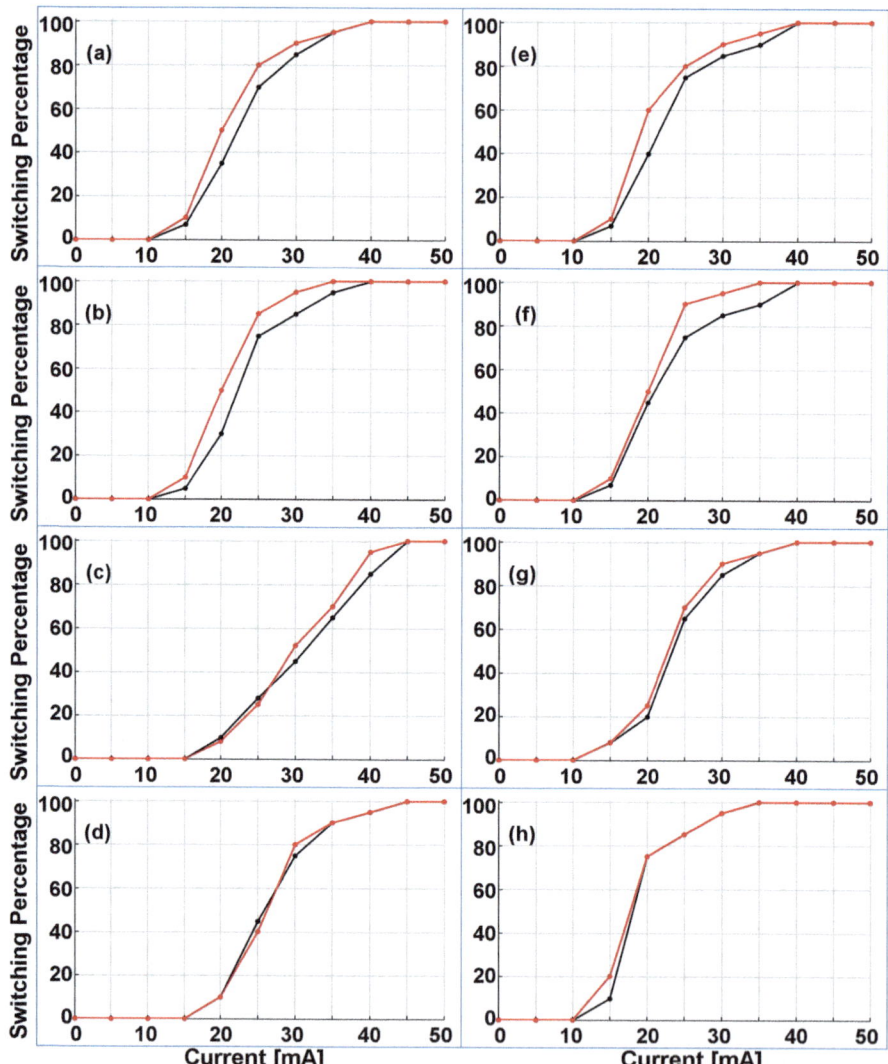

Fig. 6.19 The switching thresholds of several transistors operating in a tri-axial magnetic field are plotted. The black and red lines represent the switching thresholds for the magnetic particles with mean diameters of 8.4 μm (black lines) and 15.6 μm (red lines), respectively. These experiments are performed for the transistors shown in **a** Fig. 6.17a, **b** Fig. 6.17b, **c** Fig. 6.17c, **d** Fig. 6.17d, **e** Fig. 6.18a, **f** Fig. 6.18b, **g** Fig. 6.18c, and **h** Fig. 6.18d. The magnetic field intensity is 70 Oe with cone angle and driving frequency of 45° and 0.1 Hz, respectively. The figure is taken from [12] with permission from the Royal Society of Chemistry

The results in Fig. 6.19 show that the required switching thresholds in these transistors are relatively similar. Thus, they all are good candidates to be used in the integrated magnetophoretic circuits operating in a tri-axial magnetic field. But a tight spread of the particle trajectory shows a more reliable switching. Considering this parameter, the transistors shown in Figs. 6.17b and 6.18b seem to be the most reliable ones to be used. These two transistors are also chosen here to be further studied.

In Fig. 6.20, the switching efficiency for the transistors shown in Figs. 6.17b and 6.18b are plotted as a function of the magnetic field frequency (Fig. 6.20a, d), magnetic field strength (Fig. 6.20b, e), and the field cone angle (Fig. 6.20c, f). Figure 6.20a–f shows the analysis results for the attractive transistor presented in Fig. 6.17b and the repulsive transistor illustrated in Fig. 6.18b, respectively. Based on the reported results, the switching thresholds weakly depend on the field strengths in the range of 50–90 Oe, cone angles in the range of 37°–65°, and the driving frequencies in the range of 0.1–0.6 Hz. In these experiments, magnetic particles with a mean diameter of 8.4 μm are used.

In addition to the magnetic particles, these transistors can also switch the trajectory of the magnetically labeled single cells. In Fig. 6.21, the trajectories of magnetically labeled CD4+ human T cells in repeated experiments are shown with the overlaid blue dotted lines. The close spread of these trajectories depicts the switching reliability. Hence, this transistor can switch the trajectory of both magnetic particles and magnetically labeled cells in a similar manner.

6.4 Bends

Considering the previously introduced $\beta = r_P/r_G$ and the new parameter $\gamma = r_P/N$, where N stands for the pattern neck size in the drop-shape geometry, appropriate particle transport is seen for $\beta > 0.07$ and $\gamma < 0.07$.

Similar to the discussion for the bends for the magnetophoretic circuits operating in 2D magnetic fields, obtuse and acute bends for the circuits operating in tri-axial fields are needed [13]. A sample bend design is illustrated in Fig. 6.22, which is designed based on enlarging the gap between the drop-shape magnet numbers 1 and 2 (and 2 and 3). But this change in design leads to smaller β, which lowers the device's efficiency in transporting small particles. The bend is supposed to transport the particle from magnet 1 to magnet 3 via magnet 2. Assuming that the particle is initially at point p in Fig. 6.22a with low energies (blue area), by rotating the magnetic field in Fig. 6.22b, it has to choose to go to the low energy areas of q_1 or q_2. The magnetic energy along lines pq_1 and pq_2 at heights of 2.5 and 10 μm in Fig. 6.22c, d, allows us to realize the particle trajectory. In this plot, an energy barrier along the pq_2 path exists, which prevents the particle from moving along that path. Thus, the particle chooses the pq_1 path and cannot move from magnet number 2 to magnet number 3. The blue and red dotted lines in the inset of Fig. 6.22c show the experimental trajectories of two sample particles. The particle with the red trajectory has initially been placed on the drop-shape magnet at the corner, and, as opposed to

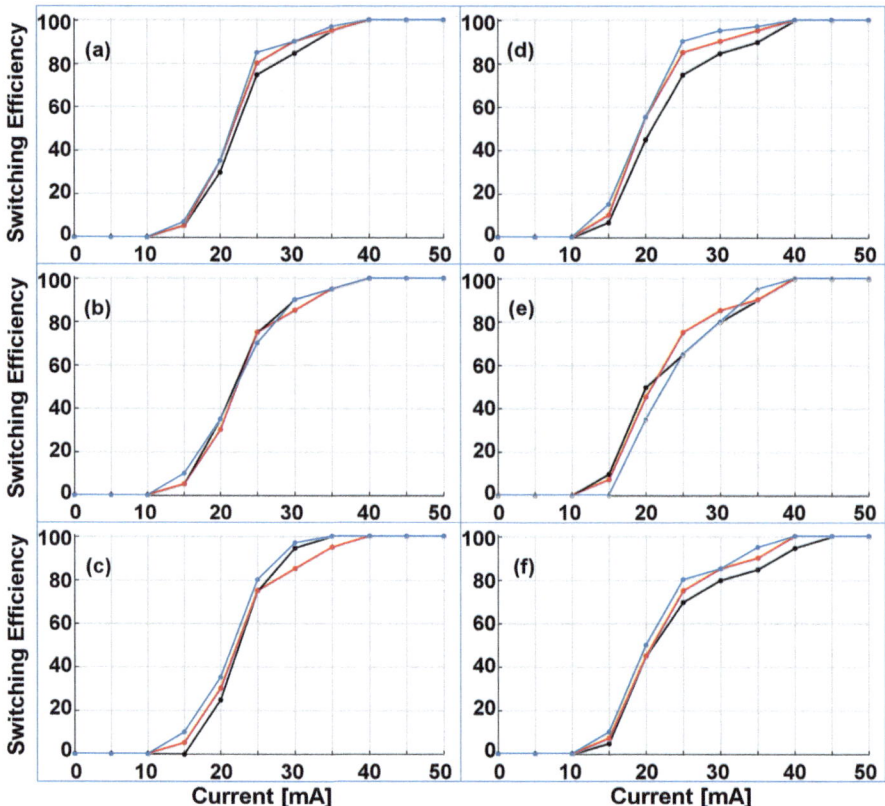

Fig. 6.20 Transistor switching efficiencies as a function of the magnetic field strength, frequency, and cone angle in a tri-axial magnetic field are illustrated. The switching efficiency is shown for the transistor represented in Fig. 6.17b for (**a**) the driving frequencies of 0.1 Hz (black), 0.3 Hz (red), and 0.6 Hz (blue) when the magnetic field strength and cone angle are 70 Oe and 45°, respectively, (**b**) the field magnitudes of 50 Oe (black), 70 Oe (red), and 90 Oe (blue) when the cone angle and the driving frequency are 45° and 0.1 Hz, respectively, and (**c**) the cone angles of 37° (black), 45° (red), and 65° (blue) with magnetic field magnitude and frequency of 70 Oe and 0.1 Hz, respectively. The results for the transistor shown in Fig. 6.18b are plotted in (**d–f**). The mean diameter of the magnetic particles in these experiments is 8.4 μm. The figure is taken from [12] with permission from the Royal Society of Chemistry

moving along the magnetic track, circulates that magnet. The particle with the blue trajectory moves from one drop-shape magnet to the next one in the straight track with appropriate β. But, then at the corner, due to an inappropriate β, it cannot move to the next drop-shape magnet and instead circulates its magnet. Thus, this is not a good bend design and cannot transport the particles on the desired paths.

The problem in the design in Fig. 6.23 arises from the fact that the disks in the corner are too far. To overcome the challenge, other designs with appropriate β and γ are needed. In the design shown in Fig. 6.23, appropriate β is achieved by introducing

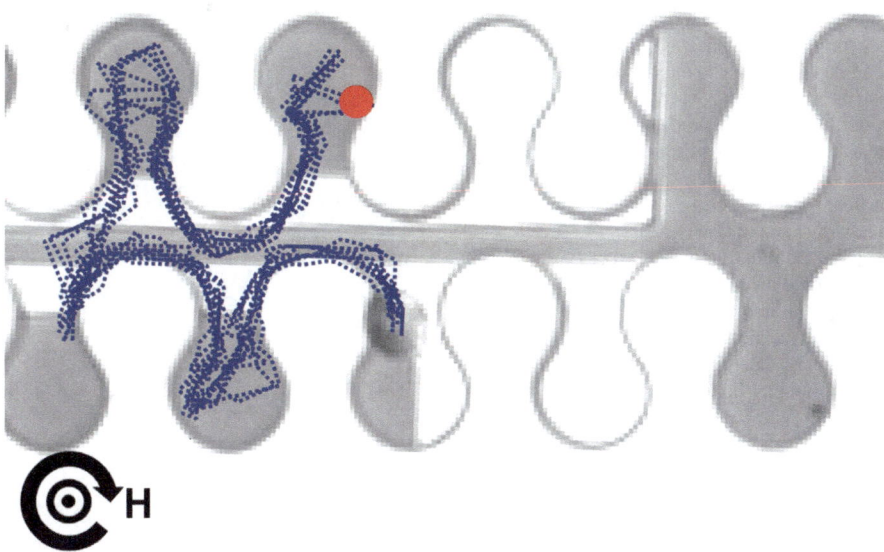

Fig. 6.21 The trajectories of the magnetically labeled human CD4+ T cells on a magnetophoretic transistor operating in a tri-axial magnetic field are shown. The applied magnetic field strength is 50 Oe with a cone angle and driving frequency of 45° and 0.1 Hz, respectively. The magnetic field rotates clockwise, and the electrical gate current is 50 mA. The red circle, the dotted blue lines, the circular arrow, and the dot at the center depict the start point of the trajectories, the trajectories of the cells, the magnetic field rotation, and the vertical bias field, respectively. The figure is taken from [12] with permission from the Royal Society of Chemistry

gaps very close to the ones in the original straight tracks shown in Fig. 6.23. In other words, as opposed to using a single magnet at the corner, many magnets, each of which has a slight angle with respect to the previous one, are used. The energy landscape illustrated in Fig. 6.23b shows how in the presented field angle (the black arrow) the particle moves. Based on this energy simulation, the energies along paths pq_1 (backward path) and pq_2 (forward path) are plotted in Fig. 6.23c, d, with the red and black curves, respectively, for two particle sizes. The energy barrier (peak) in the red curves and the negative slopes of the black curves indicate appropriate particle transport along the magnetic track. The experimental particle trajectory in Fig. 6.23a confirms this finding. Thus, this is a good design to be used as a bend. But, as a drawback, it needs a relatively large surface area on the chip.

The bends shown in Figs. 6.24 and 6.25 are two other examples that occupy less surface area on the chip. In these designs, as opposed to several drop-shape magnets a single large magnet is used. The simulation curves in Figs. 6.24c, d and 6.25c, d show that an energy barrier exists along the backward path (i.e., the red curves) which prevents the particles from moving backward. The negative energy slope along the forward path allows the particles to move in the desired direction. It is shown that the two bend designs can transport particles with a wide diameter range.

Fig. 6.22 An example of an incorrect magnetophoretic bend design. **a, b** The energy distributions for two important magnetic field angles (shown by the little black arrows) in which particle switching between the drop-shape magnets happens are illustrated. The blue and red areas depict the regions with low and high magnetic energies, respectively. Energies along the pq_1 and pq_2 lines are illustrated with the red and black curves for particles with diameters of **c** 5 μm and **d** 20 μm. The blue and red dotted lines in the inset depict the experimental trajectories for two sample particles. Reprinted from [13] under a Creative Commons Attribution 4.0 International License. http://creativecommons.org/licenses/by/4.0/

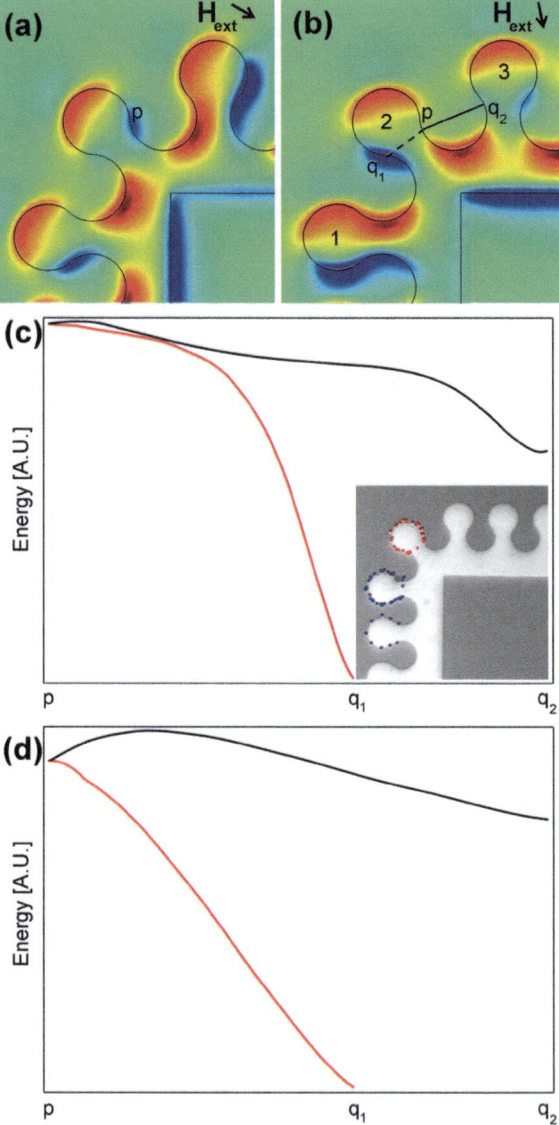

The designs discussed above are convex bends. Note that since the particle only moves on one side of the drop-shape magnetic tracks, in addition to the convex bends, concave bends are also needed. A concave bend design is shown in Fig. 6.26. The energy simulations show an energy barrier in the backward path and a negative energy slope in the forward path (see Fig. 6.26c, d). Thus, the design transports the particles on the desired path.

Fig. 6.23 An example of an appropriate magnetophoretic bend design. **a** The experimental microscopy image with overlaid particle trajectory (red dotted line) is shown. **b** The energy landscape in an important magnetic field direction (shown with the black arrow) in which particle switching between the drop-shape magnets happens is shown. The blue and red areas show the regions with low and high energies, respectively. The black arrow shows the direction of the in-plane magnetic field. A vertical bias magnetic field is superimposed to this field. Energies along the pq_1 and pq_2 lines are illustrated with the red and black curves for particles with diameters of **c** 5 μm and **d** 20 μm. Reprinted from [13] under a Creative Commons Attribution 4.0 International License. http://creativecomm ons.org/licenses/by/4.0/

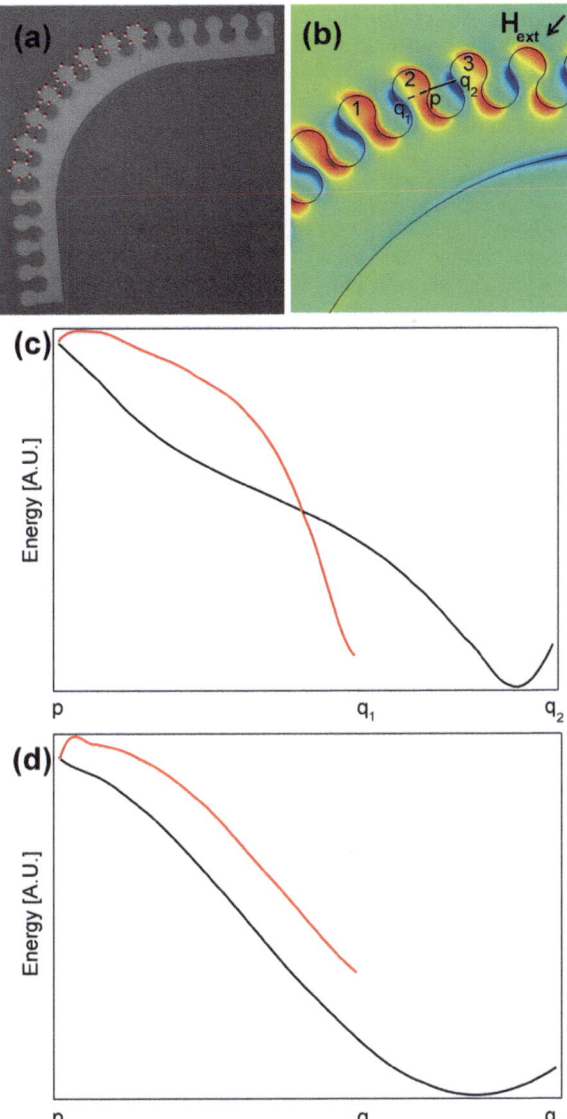

To conclude, in electrical circuits, the electrons have similar charges and thus repel each other. This repulsive force plays important roles, such as self-limiting the charge accumulation within the circuits. Hence, in the new version of the magnetophoretic circuits, to make them more similar to the electrical circuits, a vertical bias field is superimposed on the rotating field. This vertical component biases the magnetic particles such that they repel each other. Also, by eliminating one of the energy wells rotating around the magnets, the vertical bias field eradicates the degeneracy in the

Fig. 6.24 An example of a proper magnetophoretic bend design. **a** The experimental microscopy image with overlaid particle trajectory (red dotted line) is shown. **b** The energy landscape in an important magnetic field direction (shown with the black arrow) in which particle switching between the drop-shape magnets happens is shown. The blue and red areas show the regions with low and high energies, respectively. The black arrow shows the direction of the in-plane magnetic field direction. A vertical bias magnetic field is superimposed. Energies along the pq_1 and pq_2 lines are illustrated with the red and black curves for particles with diameters of **c** 5 μm and **d** 20 μm. Reprinted from [13] under a Creative Commons Attribution 4.0 International License. http://creativecommons.org/licenses/by/4.0/

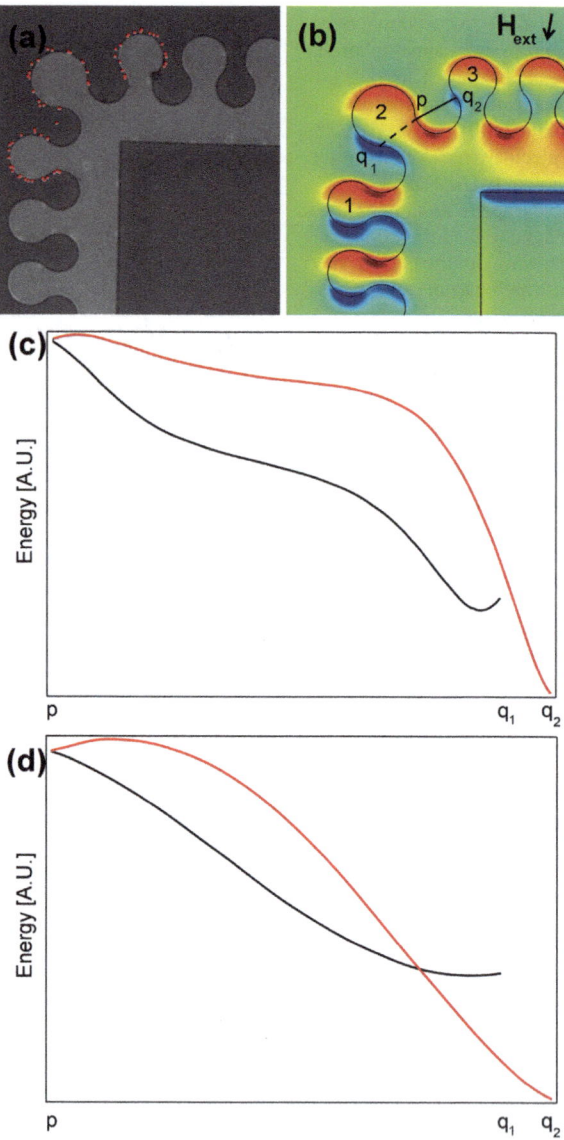

general driving clock cycle. But, the magnetophoretic circuits that operate in an in-plane rotating field cannot transport the particles in a tri-axial field. Thus, new circuit elements suitable for this magnetic field configuration are designed.

Similar to the circuits operating in the 2D field, conductors, bends, and diodes, operating in the tri-axial field were investigated. In addition to the passive circuit elements, several types of magnetophoretic transistors capable of switching the trajectory of the magnetic particles in a tri-axial magnetic field are designed. The

Fig. 6.25 An example of a proper magnetophoretic bend design. **a, b** The energy distributions for two important magnetic field angles (shown by the little black arrows) in which particle switching between the drop-shape magnets happens are illustrated. The blue and red areas depict the regions with low and high magnetic energies, respectively. Energies along the pq_1 and pq_2 lines are illustrated with the red and black curves for particles with diameters of **c** 5 μm and **d** 20 μm. The red dotted line in the inset depicts the experimental particle trajectory. Reprinted from [13] under a Creative Commons Attribution 4.0 International License. http://creativecommons.org/licenses/by/4.0/

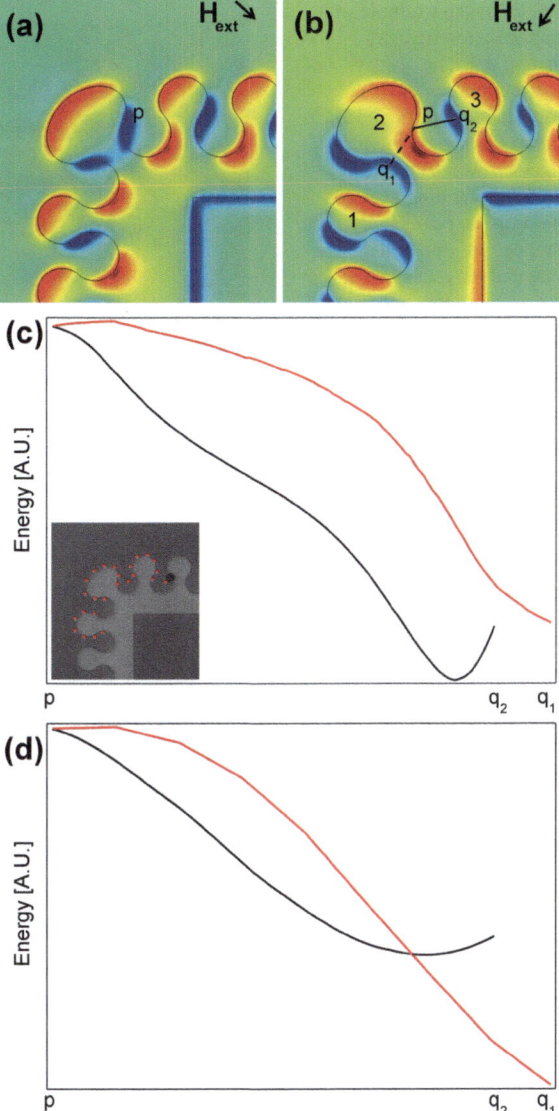

required gate currents for switching the particles and their efficiencies are studied. The magnetophoretic transistors operating in 3D fields are categorized into two main groups of attractive and repulsive transistors. Some repulsive transistors have only a repulsive gate, while in others the gate is folded to help in transferring the particle from one magnetic track to another. Because of the asymmetric nature of the energy distribution which arises from the introduction of the vertical bias field, the designs of these transistors are more complicated.

Fig. 6.26 An example of a magnetophoretic bend design. **a, b** The energy distributions for two important magnetic field angles (shown by the little black arrows) in which particle switching between the drop-shape magnets happens are illustrated. The blue and red areas depict the regions with low and high magnetic energies, respectively. Energies along the pq_1 and pq_2 lines are illustrated with the red and black curves for particles with diameters of **c** 5 μm and **d** 20 μm. The red dotted line in the inset depicts the particle experimental trajectory. Reprinted from [13] under a Creative Commons Attribution 4.0 International License. http://creativecommons.org/licenses/by/4.0/

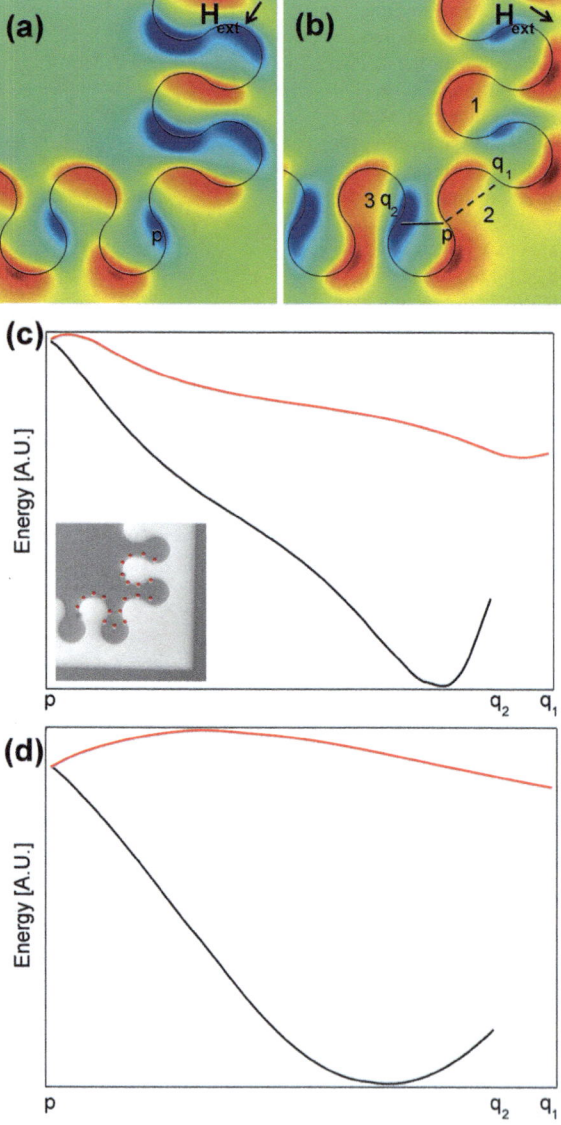

The magnetophoretic transistors operating in 2D and 3D magnetic fields require electrical gate currents of ~30 and ~50 mA, respectively. The effect of parameters including the particle size, the driving frequency, the magnetic field strength, and the magnetic field cone angle in the case of 3D fields are shown not to be challenging.

References

1. Abedini-Nassab, R., et al. (2016). Magnetophoretic conductors and diodes in a 3D magnetic field. *Advanced Functional Materials, 26*(22), 4026–4034.
2. Abedini-Nassab, R. (2022) Magnetometamaterials: Metamaterials with tunable magnetic matter conductivity. *Physical Review Applied, 17*(1), 014020.
3. Eschenfelder, A. H. (1981). *Magnetic bubble technology*. Springer.
4. Cohen, M. S., & Hsu, C. (1975). The frontiers of magnetic bubble technology. *Proceedings of the IEEE, 63*(8), 1196–1206.
5. Bobeck, A. H., Bonyhard, P. I., & Geusic, J. E. (1975). Magnetic bubbles—An emerging new memory technology. *Proceedings of the IEEE, 63*(8), 1176–1195.
6. Hsu, C., et al. (1972). A self-contained magnetic bubble-domain memory chip. *IEEE Transactions on Magnetics, 8*(2), 214–222.
7. Gergis, I., George, P., & Kobayashi, T. (1976). Gap tolerant bubble propagation circuit. *IEEE Transactions on Magnetics, 12*(6), 651–653.
8. Dell, T. H. O. (1986). Magnetic bubble domain devices. *Reports on Progress in Physics, 49*(5), 589.
9. Katsikis, G., Cybulski, J. C., & Prakash, M. (2015) Synchronous universal droplet logic and control. *Nature Physics, 11*(7), 588–596.
10. Abedini-Nassab, R., & Bahrami, S. (2021). Synchronous control of magnetic particles and magnetized cells in a tri-axial magnetic field. Lab on a Chip, *21*, 1998–2007.
11. Katsikis, G., et al. (2018). Synchronous magnetic control of water droplets in bulk ferrofluid. *Soft Matter, 14*(5), 681–692.
12. Abedini-Nassab, R., & Shourabi, R. (2022). High-throughput precise particle transport at single-particle resolution in a three-dimensional magnetic field for highly sensitive bio-detection. *Science and Reports, 12*(1), 6380.
13. Abedini-Nassab, R. et al. (2016). Magnetophoretic transistors in a tri-axial magnetic field. *Lab on a Chip, 16*(21), 4181–4188.

Chapter 7
Magnetomicrofluidic Circuits

It is reported that Gordon Moore, the co-founder and Chairman Emeritus of Intel Corporation, said in 1998 "If the auto industry advanced as rapidly as the semi-conductor industry, a Rolls Royce would get half a million miles per gallon, and it would be cheaper to throw it away than to park it [1–3]." Electronic integrated circuits (ICs), composed of circuit elements, are the important outcomes of the semi-conductor industry. Similarly, the magnetophoretic circuit elements, including the conductors, diodes, capacitors, and transistors, are combined to precisely define the trajectory of the magnetic particles and magnetized cells. The resulting integrated circuits offer the same scalability found in electronic circuits. An important example of these magnetophoretic integrated circuits is the magnetophoretic random access memory which is used to form large arrays of single particles and cells for further studies. This IC will be explained here.

7.1 Magnetophoretic Random Access Memory

The random access memories (RAM) in computers store electronic data in address-able memory cells in an array, and then later they can read the data of interest by addressing that memory cell. Drawing inspiration from these memories, magne-tophoretic RAM circuits are also developed to store single magnetic particles and magnetized cells in addressable locations [4]. Similar memory architecture is also used to address individual storage sites. In this method, the gate of the transistors corresponding to the row (word line) and column (bit line) of the target storage sites are activated. Thus, the magnetophoretic circuits send the particle to the target storage site (*i.e.*, write it) or retrieve it from that site (*i.e.*, read it). It will be discussed in more detail later, but after loading this memory with single cells, their phenotypic behavior over time can be studied. Also, it is possible to retrieve a specific target cell for follow-up genomic analysis.

R. Abedini-Nassab, *Magnetomicrofluidic Circuits for Single-Bioparticle Transport*, https://doi.org/10.1007/978-981-99-1702-0_7

The memory architecture and multiplexed addressing technique represented here have been used for years in electronic memories to decrease the wiring and signaling complexity. Similar advantages also are inherited here. The number of the needed wires to send signals to the gate of the transistors in storing/retrieving the single particles and cells is dramatically dropped. It is also possible to further drop this complexity by borrowing the concept of memory banks in computer systems, wherein several arrays form an addressable bank.

To explain the particle loading (writing) and retrieving (reading) process in the magnetophoretic RAM, a small section of it is shown in Fig. 7.1. In this chip, both attractive and repulsive modes of the transistors are used to enable the writing and reading capabilities to achieve complete memory operation. The goal is to load (write) the particles into the array and form a particle array for later studies, and, at a later time, retrieve (read) the particles of interest from the array. In Fig. 7.1, the three red lines show the trajectories for three magnetic particles being written to the storage sites numbered 42, 53, and 64. In this storage site numbering, the first and second digits stand for the row and columns, respectively. The green lines stand for the trajectories of magnetic particles being retrieved from the storage sites.

In this design, for writing the single particles into the storage sites of interest, the row and column transistors corresponding to that specific storage sites are activated in repulsive mode at the right time (*i.e.*, when the particle is on the transistor gate). For example, when the particle in row number 4 (the top row in Fig. 7.1) approaches the row transistor on the left side of that row, this transistor is switched on. Then, all transistors are kept off. When the particle moves on the gate of transistor number 42, this transistor is switched on to repel the particle and move it into storage site 42.

To retrieve the particles from the storage sites, the same column transistor used in the writing time is triggered, except it is biased in the attractive mode. Thus, this time, the particle is attracted towards the gate, which is located outside the capacitor area (See, for example, the transistor numbered 42 in Fig. 7.1). After being attracted by the transistor gate, the particle movies along the magnetic track with trajectory shown in green in Fig. 7.1.

In total, assembling the three magnetic particles and retrieving them, which is shown in Fig. 7.1, requires nine synchronized switch operations. It includes three row switches (word lines) to introduce the particles down to specific rows in the array. It also includes three column switches (bit lines), first in the repulsive mode to write the particle and then in attractive mode to read them from their storage sites. The experimental magnetic particle and magnetic cell trajectories can be seen at http://onlinelibrary.wiley.com/store/10.1002/adma.201502352/asset/supinfo/adm a201502352-sup-0006-S6.mov?v=1&s=714f1edc02ab177aef682ceb920765eff04c 3ec9 and http://onlinelibrary.wiley.com/store/10.1002/adma.201502352/asset/sup info/adma201502352-sup-0007-S7.mov?v=1&s=8a5edb54041608036d46e009794 10c04e57ab368 from [4], respectively.

Another example is shown in Fig. 7.2, where a 3×3 array is composed of other types of capacitors and transistors [5]. Figure 7.2a demonstrates how by the multiplexing system the particles can be sent to the desired capacitors. For example, when the top row transistor and the right column transistor are on, and the others are

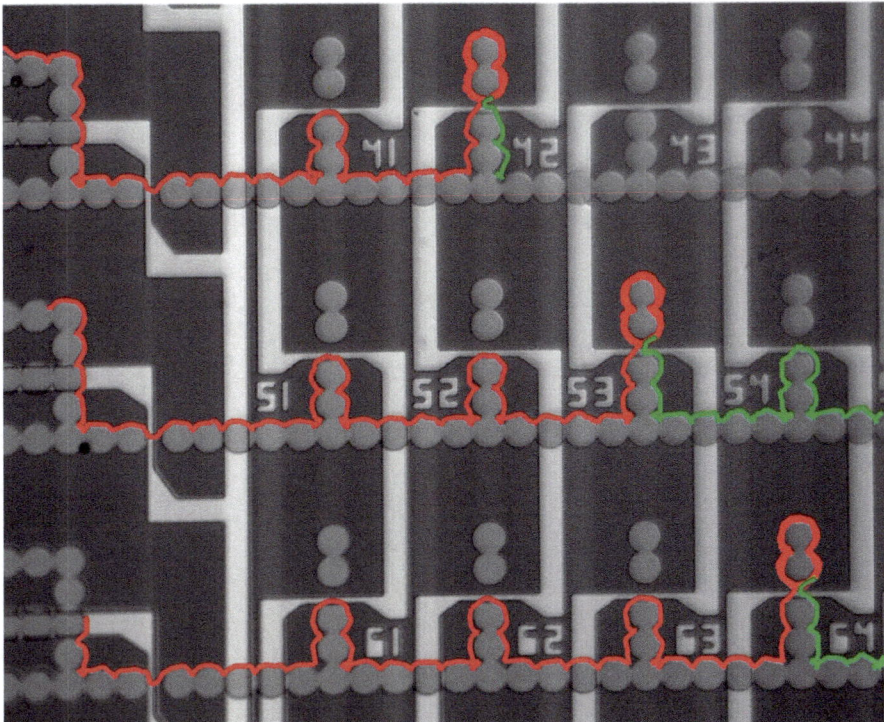

Fig. 7.1 A small section of a magnetophoretic random access memory for writing/reading magnetic particles into an array is shown. Three magnetic particles are stored in storage sites 42, 53, and 64 with the trajectories depicted in red. Then, the magnetic particles are retrieved from their storage sites with the trajectories depicted in green. The figure is taken from my [4] with permission

kept off, the particles will move along the top row and then store in the upper-right transistor (See Fig. 7.2**a**). Using similar processes and by turning on the appropriate row and column transistors in Fig. 7.2**b** and **c**, single particles are stored in capacitors 1–5. Note that in the design shown in Fig. 7.1, the particles or cells stored in the capacitors can then be retrieved (See green lines in Fig. 7.1). This achievement is due to using transistors able to operate in both attractive and repulsive modes. However, in the design shown in Fig. 7.2, this capability is not offered. That means the stored particles cannot be retrieved.

In addition to the magnetic particles, human CD4+ T cells also are reported to be stored in magnetophoretic RAMs [6]. A similar addressing protocol is needed for turning on and off the required transistors and storing the cells at desired capacitors.

Fig. 7.2 Another example of RAM is shown. (**a–c**) Time sequences of particle loading are illustrated. Numbers 1–5 show the capacitors in which particles are stored. The white arrows and the red circular arrow depict the particle trajectories and the external magnetic field rotation, respectively. The figure is taken from [5] under a Creative Commons Attribution 3.0 Unported License. http://creativecommons.org/licenses/by/3.0/

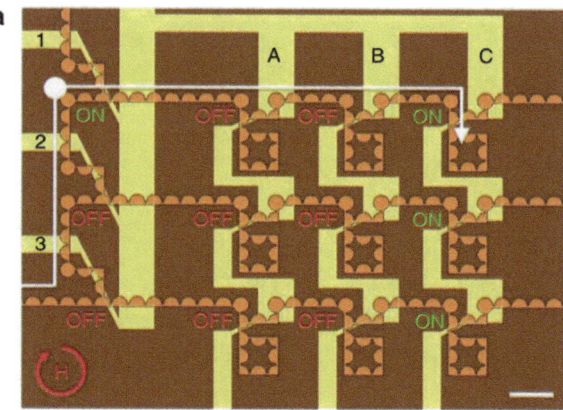

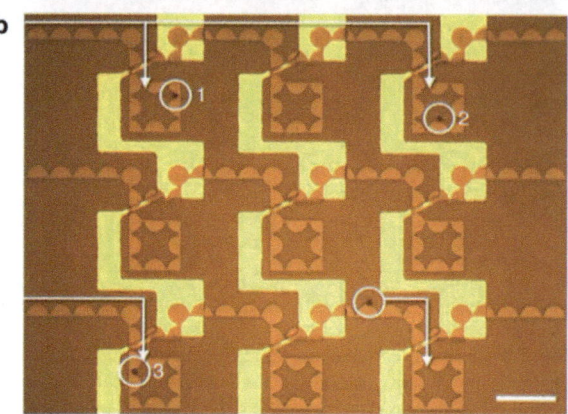

7.2 Magnetomicrofluidics

As explained in the last subchapter, magnetophoretic circuits can precisely manipulate the magnetic particles and magnetized cells in parallel. In these circuits, the particles move based on magnetic forces, which can be pre-programmed and locally addressed. The magnetic forces are gentle and can be removed when they are not needed. All these specifications are positive and make these chips good candidates to be used in bio-applications.

Assuming that a magnetophoretic circuit operates at a driving frequency of − 0.5 Hz and its conductor period is 20 μm, this chip moves the particles 10 μm per second in the desired direction. But in large arrays, the particles may travel distances as large as a centimeter. That means this system is slow for populating an array in that size range. To overcome this problem, hybrid systems are developed.

One good solution for this challenge is to combine the magnetophoretic circuits with microfluidic channels. The microfluidic channels move the particles with hydrodynamic forces at much higher speeds compared to the magnetophoretic circuits. But they do not offer precise control over individual particles. This important capability is provided by the magnetophoretic circuits. But another mechanism is needed to switch the particles from the fluid flow to the magnetic forces. This requirement is answered by adapting the hydrodynamic traps discussed in Chap. 1.1.2 Array-Based Systems.

The combination of the magnetophoretic circuits with hydrodynamic trapping systems results in magnetomicrofluidic chips. In this technique, the particles are first transported using the hydrodynamic forces in the microfluidic chips. Then, individual particles are captured by the hydrodynamic traps. Once captured, magnetophoretic circuits are used to transfer them to the storage sites.

But why do we need to transfer the trapped particles to the storage sites? What are the advantages of this method over pure hydrodynamic trapping systems? The answer is: (i) In the trap sites, strong shear stress may affect the cell behavior or viability [7]. But, by moving them to the storage sites, they do not face this problem anymore. (ii) By repeating the protocol mentioned above, the second set of particles can also be introduced to the array. Thus, it is possible to form cell pairs to study the cell–cell interactions. (iii) The storage sites are isolated microchambers. But the trap sites are not isolated, and the particles shed from a cell from another trap may move to another cell and alter the single-cell results.

Designing a hydrodynamic trapping microfluidic system is explained in Chap. 1.1.2 Array-Based Systems. Specially designed microfluidic channels that distribute the flow patterns in a resistive laminar flow network are used to trap the single particles and cells in an array (See Fig. 7.3**a**). This particle trapping system is designed by including a trap fluidic pathway in parallel to a bypass pathway in the microfluidic network. The hydrodynamic resistivity of the trap pathway (shown with index 1 in Fig. 7.3) is lower initially compared to the bypass pathway (shown with index 2 in Fig. 7.3). The equivalent circuit model is shown in Fig. 7.3**b**. Equation (7.1) introduced in Chap. 1.1.2 Array-Based Systems and copied here, can be

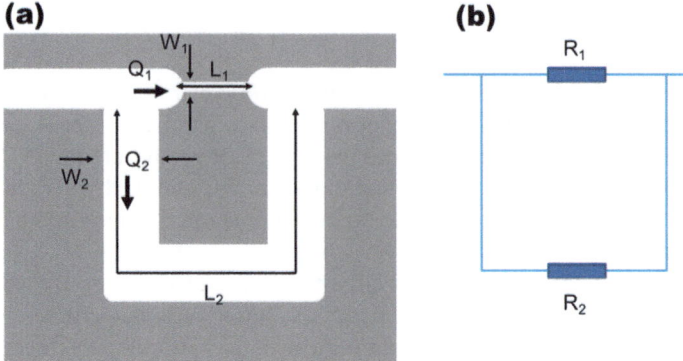

Fig. 7.3 A schematic of a hydrodynamic trapping system and the equivalent circuit model for this system are shown. (**a**) A hydrodynamic trapping system composed of two parallel fluidic paths is illustrated. Here index 1 stands for the trap path, while index 2 stands for the bypass path. (**b**) The equivalent circuit model for the system presented in panel **a** is shown. R stands for fluidic resistivity

used to calculate the flow rates [8].

$$\frac{Q_t}{Q_b} = \left(\frac{c_b(\iota)}{c_t(\iota)}\right)\left(\frac{L_b}{L_t}\right)\left(\frac{W_b + H_t}{W_t + H_b}\right)^2\left(\frac{W_t H_t}{W_b H_b}\right)^3. \tag{7.1}$$

Because the flow rate in the trap pathway compared to the bypass pathway is initially larger, the particle tends to move to that site. Since the particle size is larger than the trap size, it gets trapped. After trapping, the flow rate in this channel drops, and the next particles move through the bypass channel, leaving single particles in the trap sites. Note that the trap width (*i.e.*, W_1) must be smaller than the diameter of the particles. Also, the channel heights are equal, so the bypass pathway must be longer than the trap pathway (*i.e.*, $L_2 \gg L_1$) to obtain the required ratio of the flow rates.

One important parameter to be considered in the particle trapping system is the maximum possible pressure to be applied to the cells. The cell membrane is flexible, which means the cells are not rigid. Thus, when captured in the trap site, under extra force they may squeeze through the trap site and escape. This problem not only ruins single-cell array formation but also may affect cell viability and behavior.

Additionally, when the cells are captured in the trap sites, an extra flow rate may result in strong shear stress, even if they do not escape the trap site. Thus, in the cell trapping experiments, applying gentle pressures to the inlet and ensuring the right flow rates are crucially important.

So, the microfluidic channels transport the particles using the hydrodynamic forces, and then the hydrodynamic trapping system captures the single particles in the trap sites. This process, as discussed in the previous subchapters, forms the first phase. But a complete magnetomicrofluidic circuit, in addition to the microfluidic part used in the first phase, is composed of the magnetophoretic circuits, which are used in the second phase. In this phase, the magnetophoretic circuits send the trapped particles from the trap sites into the storage sites [9].

During the first phase, the magnetic fields are kept off. This strategy is important in order to prevent the transport of the trapped particles from the fluidic trap sites at this phase, which will result in capturing the next approaching particle in the same trap site and ultimately having more than a single particle in a single storage site. Only after the first phase is complete and all the traps are filled with the single particles, the magnetic fields are turned on and the second phase begins.

Figure 7.4 illustrates an example of magnetomicrofluidic circuits, in which the PDMS-based microfluidic channels are bonded on top of a magnetophoretic chip. In this figure, the two mentioned phases are depicted. First, in Fig. 7.4a, the particles are shown in the trap site after the first phase. Then, as shown in Fig. 7.4b magnetophoretic circuits are used to transfer the captured single particles from the trap sites into the storage sites. The dotted lines in this figure stand for the trajectory of a particle, which is shown in the red circle. The black arrows depict the direction of the particle transfer.

The chip shown in Fig. 7.4 consists of a 30×20 storage site array. The experiment is repeated for magnetic particles with size ranges of 5–5.9 µm (called 5 µm in Fig. 7.4), 10–13.9 µm (called 10 µm in Fig. 7.4), and 14–17.9 µm (called 15 µm in Fig. 7.4). Some trap sites may remain empty, and some may capture two beads. Also, for some particles, transporting them from the trap sites into the storage sites may be challenging. The statistical results for the number of particles in trap sites and the storage sites are plotted in Fig. 7.4c and d, respectively. Based on these results, overall particle trapping efficiency in the first phase is ~80%. As high as 85% of these particles are also reported to be transferred to the storage sites in the second phase.

As seen in the plots in Fig. 7.4, in some trap sites, more than a single particle are trapped. An example of this scenario is displayed in Fig. 7.5. In Fig. 7.5a, the first particle is trapped. In Fig. 7.5b, the second particle comes into the trap. To overcome this problem, a rotating magnetic field in the reverse direction (*i.e.*, counterclockwise in Fig. 7.5) can be used. The applied magnetic field creates energy minima and maxima circulating the magnetic disks. As illustrated in the inset of Fig. 7.5b, if the energy well is on the trap site, the particles remain there. But by rotating the magnetic field, the energy well moves towards the channel wall on the trap site. But since the wall prevents the particles from following the energy well, they remain close to the wall, and then the approaching energy maximum repels the trapped particle (See Fig. 7.5c and its inset).

The hydrodynamic force applied to the particles is stronger at points closer to the trap site. In the other words, the hydrodynamic force on the first trapped particle is stronger than the force on the second particle. Thus, by tuning the flow rate, the hydrodynamic force on the second particle is set to be weaker than the magnetic repulsion forces, but it is stronger for the first particle. As a result, the second particle is repelled from the trap site while the first particle remains there.

It is also possible to form arrays of particle pairs [9]. To achieve this goal, after filling the storage sites with the first particles, the same loading process explained

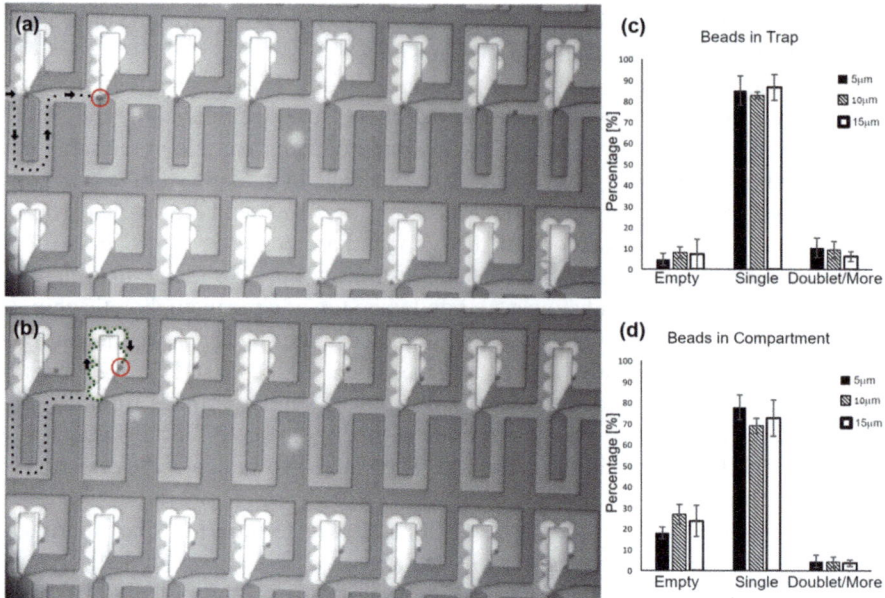

Fig. 7.4 Assembling magnetic particles in a magnetomicrofluidic chip is shown. (**a**) In the first phase, the magnetic particles are captured in the hydrodynamic traps. (**b**) The captured particles are moved into the storage sites with the magnetophoretic circuits. The trajectory of a sample magnetic particle and its movement direction are shown by the dotted lines and the black arrows, respectively. The red circle in each panel shows a sample particle position. (**a**) The particles are first trapped in the hydrodynamic trap sites. (**c** and **d**) The capture efficiency for particles with diameters of 5 μm (black bars), 10 μm (patterned bars), and 15 μm (white bars) (see text for the size ranges) in the trap sites, and the compartments are plotted in (**c**) and (**d**), respectively. © 2019 IEEE. Reprinted, with permission, from [9]

above is repeated to assemble the second particle set in the same storage sites. Figure 7.6 illustrates an example of particle pair formation using green and red fluorescent beads with diameters in the range of 7.5–8.5 μm. A small section of an overlaid fluorescent detection on an optical microscopy image is demonstrated. The particle pair loading efficiency is plotted in Fig. 7.6**b**.

The particles shown in Fig. 7.6 are magnetic beads, which are successfully assembled into the illustrated array. But transporting the magnetized cells inside the microfluidic channels with the magnetophoretic circuits is a more challenging task. Thus, the magnetomicrofluidic chips must be modified further.

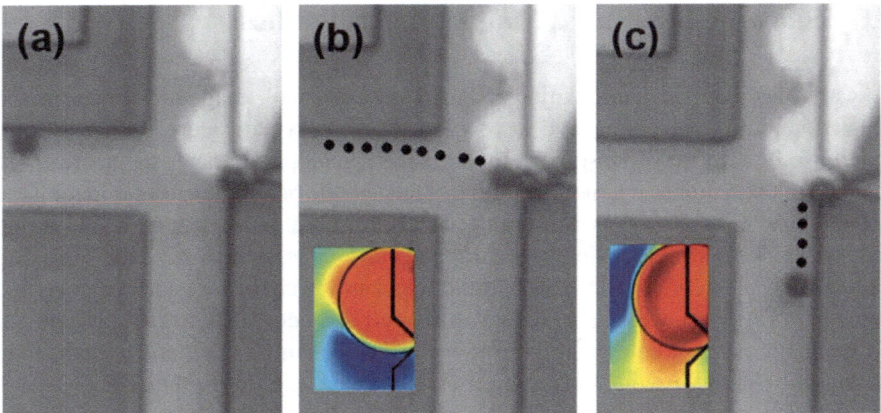

Fig. 7.5 Magnetic repelling of the extra captured particles from the trap sites is shown. (**a**) The first particle is trapped in the trap site, and the second particle is approaching. (**b**) The second particle is captured too. The simulation results in the inset show a magnetic energy minimum (blue area) on the trap site. (**c**) The second particle is repelled. The energy simulation in the inset shows an energy maximum (red area) on the trap site. The dots stand for the particle trajectories. © 2019 IEEE. Reprinted, with permission, from [9]

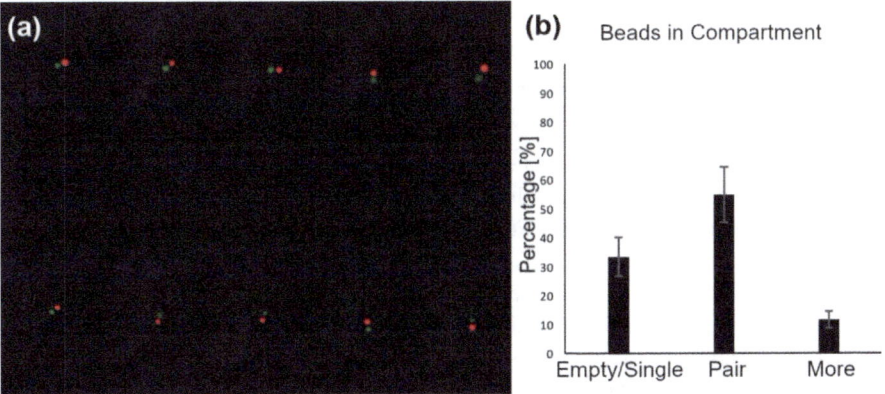

Fig. 7.6 Particle pair assembling is shown. (**a**) The overlaid fluorescent detection and optical microscopy images are shown. The green and red fluorescently labeled particles are detected. (**b**) The magnetomicrofluidic chip efficiency in assembling particle pairs in the compartments is plotted. © 2019 IEEE. Reprinted, with permission, from [9]

7.3   Silicon-Glass-Based Magnetomicrofluidics

Towards the goal of manipulating magnetized cells with magnetic forces in the magnetomicrofluidic chips, two key points need to be considered. First, during handling, the cells may lose some of their magnetic nanoparticles which may result in the lack of the required magnetic moment. Thus, gentle handling before injecting the cells into the chip as well as during injection is required. For example, it is possible that strong shear stress applied to the trapped cells leads to the loss of the magnetic nanoparticles. The magnetic nanoparticle loss can lower the cell surface coverage area, ψ, in Eq. (3.3), which also results in the cell magnetization drop. Although not all magnetic nanoparticles may be removed, the remaining magnetic content may not be sufficient to provide the required magnetic forces to move the cells.

Another point to be considered in manipulating the cells with magnetomicrofluidic chips is the chip surface chemistry. The cells have proteins on their surface, which may make bonds to the microfluidic walls. In magnetophoretic circuits (*i.e.*, open chips), chip surface passivation approaches (See Chap. 4.2.1 SiO_2 Surface and Chap. 4.2.2 SU8 Surface Passivation) provide a non-fouling layer to prevent cells from sticking to the chip surfaces. Hence, growing the polymer brushes inside microfluidic channels to lower the chance of cell-wall bond formation is needed.

7.3.1   Microchannel Surface Passivation

To grow a non-fouling layer inside magnetomicrofluidic channels and prevent the cells from adhering to the microfluidic chip surface or walls, the polymerization protocol explained in 4.2 Surface Functionalization needs to be modified. This need arises from the fact that the magnetophoretic circuits are on open chips; however, the magnetomicrofluidic chips are closed. Syringe pumps are used to flow the precursors in the liquid phase into the microfluidic channels. But, as explained before, PDMS is a typical material that is used in most microfluidic chips, similar to the chip shown in Fig. 7.4. The problem is that many solvents swell or delaminate the PDMS. Figure 7.7 illustrates a PDMS-based microfluidic chip before and after a POEGMA polymerization attempt. It is obvious that the PDMS is delaminated, and the microfluidic channels are destroyed (compare Fig. 7.7**a** with Fig. 7.7**b**). Thus, ethanol, which is the solvent in POEGMA polymerization, is not suggested for PDMS-based microfluidic chips. Other passivation methods, such as covering the chips with water-based PEG, are suggested.

To answer the mentioned problem, it is possible to create microfluidic channels in silicon, as opposed to the PDMS. In this method, the microfluidic channels are etched (using DRIE) into the silicon wafer, and then anodic bonding is used to bond the glass lid to the silicon wafer. But fabricating magnetophoretic circuits inside the etched channels is challenging. The microchannel walls prevent proper photoresist coating and lighting inside the channels. Also, they prevent the metals during the

Fig. 7.7 The effect of ethanol on a PDMS-based magnetomicrofluidic chip during polymerization is shown. The chip (**a**) before and (**b**) after the polymerization is illustrated

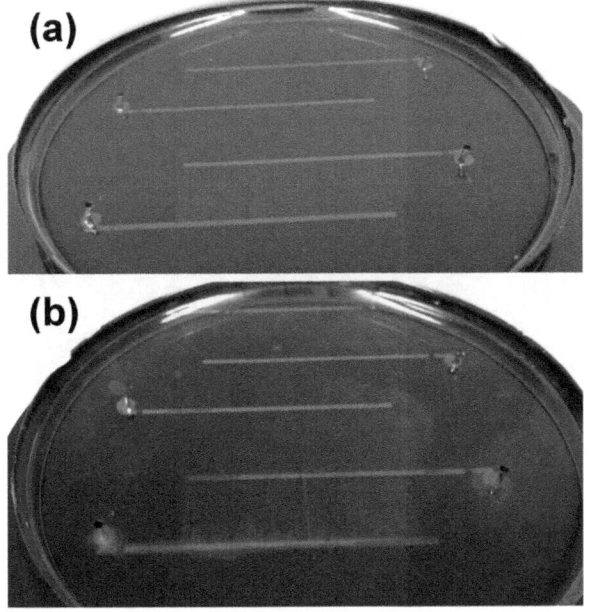

metal evaporation step when coating the chip. Furthermore, precise alignment in anodic bonding is not easy. Thus, fabricating the chips based on glass lids with magnetophoretic circuits on them and bonding them to the silicon wafers with etched microchannels is challenging.

A novel self-aligned fabrication protocol is reported to answer the challenges mentioned above. This method is based on including the magnetophoretic circuits into the microfluidic walls in a silicon substrate and then anodic bonding the silicon substrate with a glass lid. Hence, this method needs a new design for the microfluidic channels which is shown in Fig. 7.8**a**. A silicon wafer with microchannels etched into it is shown in Fig. 7.8**b**. In other words, in this method, the microfluidic structures play the role of both the microchannels and the magnetophoretic circuits. Here, the magnetic layer covers the whole chip surface and not inside the microchannels (See Chap. 7.3.2 Silicon-Glass-Based Magnetomicrofluidic Fabrication Protocol). Thus, the fabrication of magnetophoretic circuits is not challenging. Also, since there is no design on the glass lid, no challenging alignment is needed.

Since the new chip is based on silicon and glass, it can carry solvents such as ethanol. Thus, growing polymer brushes, such as POEGMA, which was challenging in PDMS-based chips, is easy in the new chips. The polymerization protocol is similar to the one mentioned in Chap. 4.2.1 SiO_2 Surface Passivation, except that the precursors are injected into the microfluidic chips in a liquid phase.

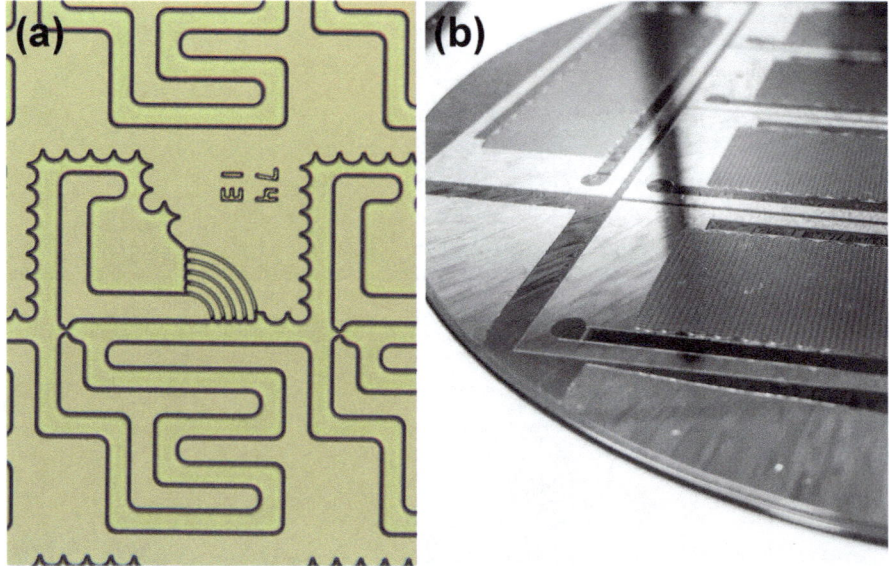

Fig. 7.8 Silicon-based magnetomicrofluidic design is shown. (**a**) The design, in which the magnetophoretic circuits are included in the walls, is shown. (**b**) A picture of an etched chip is illustrated. This picture is taken by Justin Gladman, the R&D Engineer at Shared Materials Instrumentation Facility (SMIF) at Duke University

To evaluate the flow velocities at the trap site before and after a particle is captured, FEM simulations are used [10]. The simulation results are presented in Fig. 7.9. By integrating the flux over the cross-sectional area at each channel, achieving appropriate flow rates is ensured. In other words, when no particle is trapped, $Q_{trap} > Q_{bypass} > Q_{compartment}$. But for an occupied trap site, $Q_{bypass} > Q_{trap} > Q_{compartment}$. The simulation analysis in Fig. 7.9 also shows the low flow velocities in the storage site (blue region) compared to the one in the trap site (red region). This result confirms that the shear stress in the storage sites is low compared to that of the trap sites, which is an important need for cell culture.

7.3.2 Silicon-Glass-Based Magnetomicrofluidic Fabrication Protocol

Photolithography is the first step in fabricating the chips based on silicon and glass. It is used to pattern the microfluidic channels on the silicon wafer. The detailed lithography protocol is provided in Chap. 4.1.1 Photolithography. Here, a positive photoresist is used, and then the channel is etched into the silicon wafer with DRIE. The remaining photoresist on the chip is removed by resist remover at 65 degrees C for 30 min. Then, the chip is further cleaned with acetone and isopropanol and

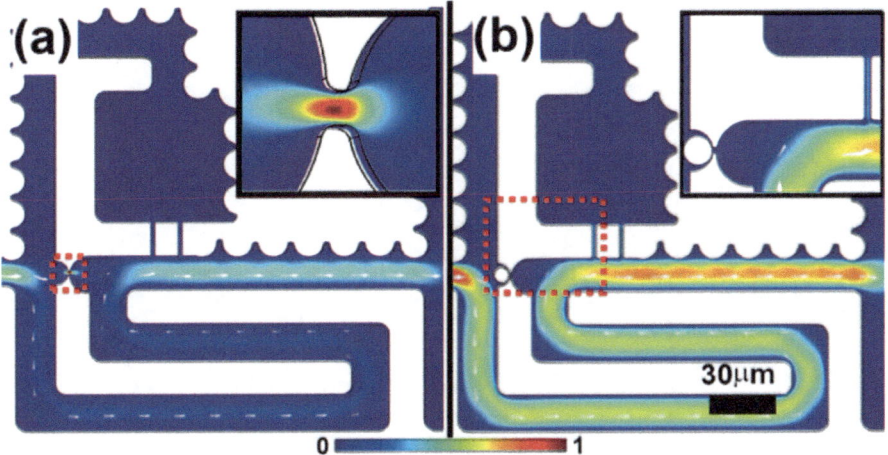

Fig. 7.9 The simulation results for the hydrodynamic flow rates in microchannels are shown. The relative flow velocities for (**a**) an empty trap site and (**b**) a trap site after capturing a particle are shown. The blue and red regions stand for the regions with low and high flow velocities. The white arrows shows the direction of the fluid flow at each point. The insets show zoomed views of the trap sites. The input flow velocity is set to 1 mL/min. © [Year] IEEE. Reprinted, with permission, from [10]

dried with nitrogen gas. A polymer is deposited on the chip during the etch process and needs to be removed by piranha solution (3:1 ratio of sulfuric acid to hydrogen peroxide) overnight, followed by a thorough rinse with water. Note that handling the piranha solution requires extra care. The chip is then dried with nitrogen. The resulting chip is shown in Fig. 7.8**b**.

To ensure the right microchannel depth is created, it can be measured with different methods. One simple way is to do that with a profilometer, which plots the channel height along a defined scanning line. A sample measurement result is illustrated in Fig. 7.10.

Next, the magnetic thin film is created. To do that, a ~100 nm-thick layer of permalloy (it can be a bit thicker, too) is deposited on the chip surface. This step can be done by a metal evaporator machine. Now, the chip has the required microchannels with magnetophoretic circuits in the walls. But the problem is that the top layer is permalloy, and thus sealing the channel would be challenging. To answer this problem, the whole chip is covered with a 200 nm-thick layer of SiO_2 using a PECVD tool. The silicon dioxide layer works well in the anodic bonding process and bonds to a glass lid. The detailed protocols for the metal evaporation and SiO_2 deposition are explained in Chap. 4.1.2 Metallic Thin Film and Chap. 4.1.4 Insulating, respectively.

To seal the chip, an anodic bonding method is used. First, inlet and outlet holes are drilled in a Borofloat glass slide. After cleaning it with acetone and isopropanol, it is dried with nitrogen gas. Then, the glass slide is placed in piranha solution (3:1 ratio of sulfuric acid to hydrogen peroxide) for 10 min. Then, it is rinsed thoroughly with water followed by a nitrogen drying step. Note that handling piranha solutions

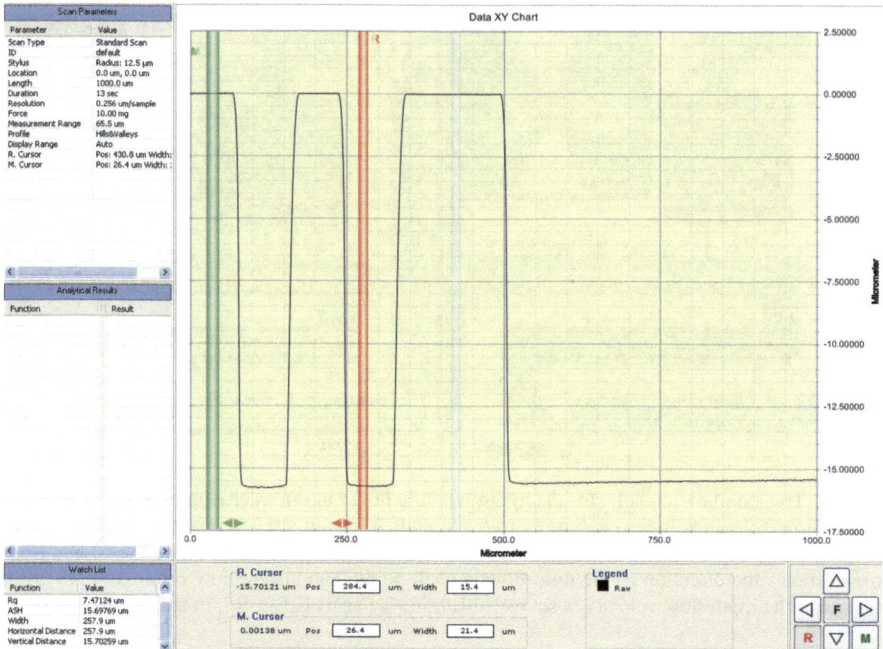

Fig. 7.10 The profilometer result for measuring the depth of a microchannel is shown. The profilometer tip moves along a defined straight line on the chip surface and measures the heights along that line. In this example, the microchannel depth is reported to be ~15.7 μm

requires extra care. Now the chip and glass slides are ready for anodic bonding. The glass piece is carefully aligned on the silicon chip and placed on a hotplate at 450° C. The two pieces are pressed together to keep them in contact. For this reason, a weight of ~ 5 kg can be placed on top of the two. A voltage of 1000 V is applied between the glass slide and the silicon chip for 2 h. In Fig. 7.11, the anodic bonding setup is shown.

7.3.3 Assembling and Interfacing

Properly connecting the chip to the world outside is an important task to use in the labs. For the magnetophoretic chips, electrical and fluidic connections are needed. Providing the electrical connection is explained in 4.3.2 Electrical Connections. The fluids are provided to the chip by the tubes. Since PDMS is elastic, in PDMS-based chips, the polymer holds the tubing in the inlet and outlet holes. But when the chip is based on rigid materials such as silicon or glass, the pipes need to be fixed in the holes with a different method.

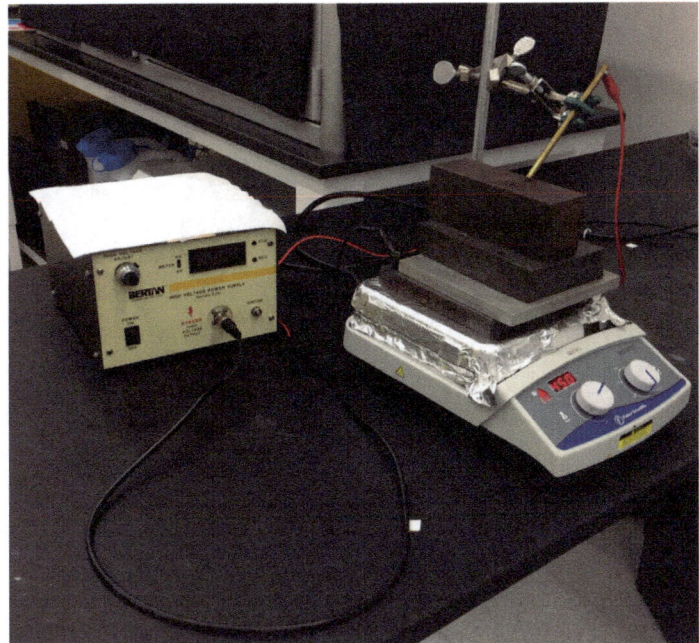

Fig. 7.11 Anodic bonding setup is shown. The glass lid and the silicon wafer are placed on the hotplate (450° C) and under a weight of ~ 5 kg. A power supply applies 1000 V to the overlaid structure

A simple method is to use plasma treatment (as explained before) to bond a small piece of PDMS, with a hole in it, to the glass on top of the inlet/outlet (See Fig. 7.12**a**). Thus, the tubing can be plugged into this PDMS slab. If polymerization ruins the PDMS slab, it can be replaced with a new one. Although it is a simple method, it is not the most suitable method.

In another simple method, Luer fittings can be epoxy glued at the inlet and outlet. This approach is illustrated in Fig. 7.12**b**. The fluid and particles (including the cells) are gently injected into the chip using a syringe pump (See Fig. 7.12**c**). Since introducing glue to the chip may affect cell viability, this method may not be the best bio-friendly solution.

Using commercially available microfluidic clips is a good approach to answer this need. In this method, the plastic ports are screwed into the clip providing the required sealing on the chip ports. In order to provide the required temperature for the cells to live in magnetomicrofluidic chips, the setup can also be equipped with a thermal controller. Now by introducing a CO_2 gas sensor/controller, this system is ready for culturing the cells without requiring a bulky incubator.

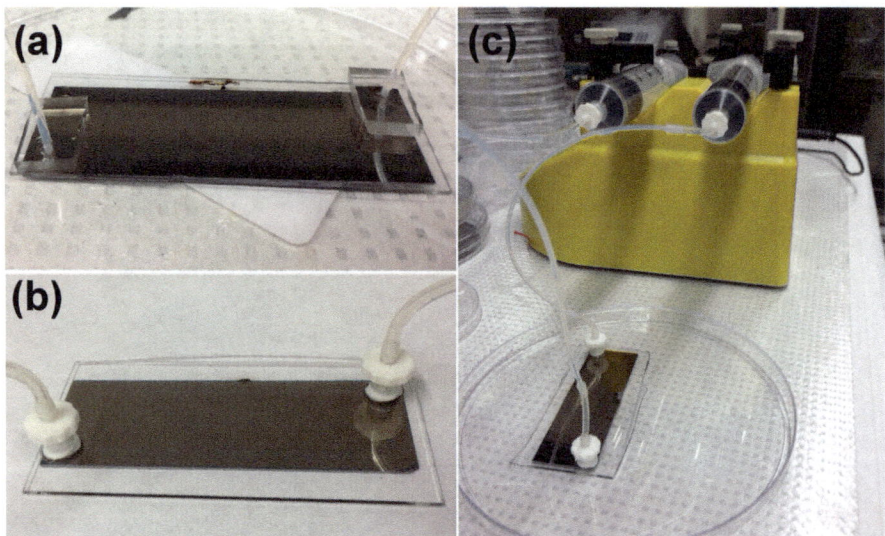

Fig. 7.12 Silicon-based magnetomicrofluidic chips are shown. The ports are connected using (**a**) plasma-treated PDMS slabs or (**b**) glued Luer fittings. (**c**) The particles and fluids are gently delivered to the chip by a syringe pump through a tubing

7.3.4 Magnetomicrofluidic Circuit Operation

The fabrication protocol of the silicon-based magnetomicrofluidic chip was discussed in Chap. 7.3.2 Silicon-Glass-Based Magnetomicrofluidic Fabrication Protocol, and it was mentioned that in this chip, the whole surface is coated with permalloy. Thus, in addition to the microchannel walls which form the magnetic circuits, inside the microchannels magnetic materials exist. That means we have two magnetic layers' distances with the microchannel heights. Hence, the effect of this extra magnetic thin film must be investigated. Towards this goal, in an experiment, a ferrofluid suspension is introduced into the chip to simply visualize the potential magnetic energy distribution (a method similar to the Bitter domain decoration) [10]. Now, the chip is exposed to a rotating magnetic field, and the ferrofluid behavior (*i.e.*, nanoparticle movement and their distribution) is evaluated. The experimental results are illustrated in Fig. 7.13a–d. The energy distributions corresponding to the experimental results of Fig. 7.13a–d are also simulated with FEM methods, the results of which are shown in Fig. 7.13e–h, respectively. The simulation and experiments agree well.

FEM energy simulations at various heights can be used to study the effect of the two magnetic layers, too [10]. Hereafter let's call the magnetic layer in the microfluidic walls and the one inside the channels the wanted magnetic film and the unwanted magnetic film, respectively (See Fig. 7.14a, b). In Fig. 7.14c, the energy distribution in a horizontal plane close to the wanted magnetic film is plotted. This energy landscape shows that appropriate energy wells are created near the

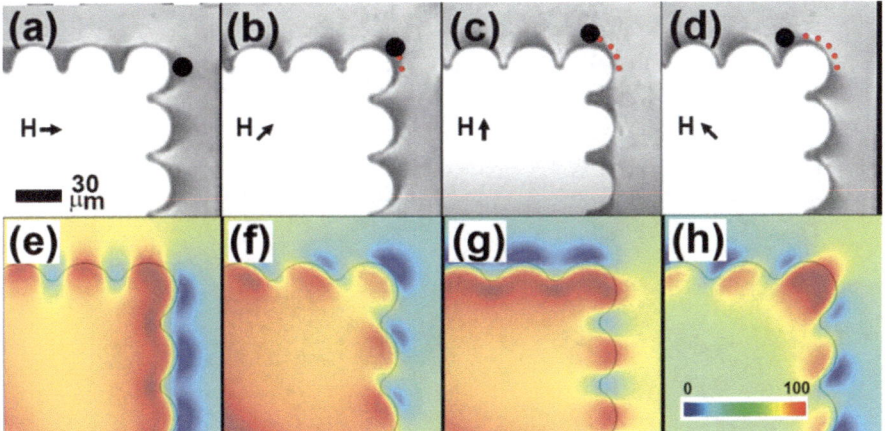

Fig. 7.13 The behavior of ferrofluid in a magnetomicrofluidic chip is shown. The distribution of the magnetic nanoparticles can visualize the energy distribution. The black arrow in each panel depicts the direction of the external magnetic field. The energy distribution evaluated by the ferrofluid distribution is used to find the position of a hypothetical magnetic particle depicted by a little black circle in each panel. The red dotted lines stand for the trajectory of that hypothetical particle. The corresponding energy distributions for the cases shown in (**a–d**) are illustrated in (**e–h**), respectively. The blue and red areas represent the regions with low and high magnetic energies, respectively. © [Year] IEEE. Reprinted, with permission, from [10]

wanted magnetic film and they trap and transfer the particles in the desired direction. However, in the energy simulation results, at heights far from the wanted magnetic film (*i.e.*, the middle of the two films (See Fig. 7.14**d**) or close to the unwanted magnetic film (See Fig. 7.14**e**) appropriate energy wells are not seen. Also, Fig. 7.14**f** illustrates the energy distribution along the black horizontal line shown in Fig. 7.14**d** at various heights. These plots confirm that the required energy wells disappear at distances more than 0.6 h from the wanted magnetic film, where h stands for the microchannel height. Thus, it is suggested to put the chip such that the glass slide is faced down and the silicon wafer is up. So, the gravity force pulls down the particles close to the desired magnetic film and desired energy distribution. The ratio of the channel height to the particle radius must follow $\frac{r_p}{h} < 0.4$, where r_p is the particle radius. To make the chip useful for a wide particle range, $\frac{r_p}{h} < 0.25$ is suggested.

The silicon-based magnetomicrofluidic circuit works similar to its previous version, based on PDMS, introduced in Chap. 7.2 Magnetomicrofluidics. But the particles may experience a strong fluidic force at the hydrodynamic trap. This strong force can overcome the magnetic forces and become challenging. Thus, in the silicon-based magnetomicrofluidic circuit design, the magnetophoretic circuits are distanced from the hydrodynamic traps (compare Fig. 7.4 with Fig. 7.8**a**), where the fluidic force is weaker.

The device operates in three phases. In the first phase, the single particles are injected into the microchannels and trapped in the hydrodynamic trap sites. This phase is similar to phase 1 in the PDMS-based magnetomicrofluicic chips which are

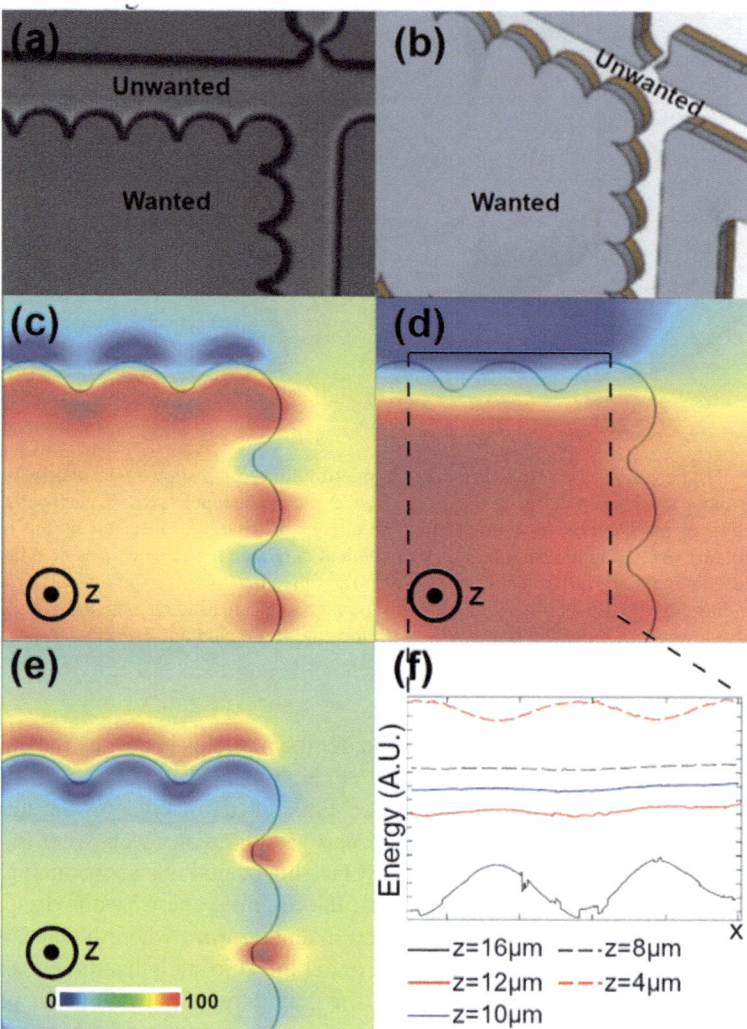

Fig. 7.14 The magnetic energy landscape at different heights in a silicon-based magnetomicroflu-idic chip is illustrated: (**a**) top and (**b**) perspective views of the chip are shown. The wanted and unwanted magnetic films are visible in these pictures. The FEM simulation results in a plane (**c**) close to the wanted magnetic film, (**d**) in between the two films, and (**e**) close to the unwanted magnetic film are illustrated. The blue and red regions stand for the spots with low and high magnetic ener-gies, respectively. (**f**) The magnetic energy along the black horizontal line in (**d**) at different heights is plotted. In the legend, z depicts the height at which the magnetic energy simulation is run. The external magnetic field strength in these simulations is 80 Oe. © [Year] IEEE. Reprinted, with permission, from [10]

explained in Chap. 7.2 Magnetomicrofluidics. After having the particles trapped, in phase 2, a reverse fluid flow is applied to move the trapped particles out of the trap sites towards the magnetophoretic circuits. The magnetic field is not turned on, and these circuits attract the magnetic particles. Next, in phase 3, the magnetic field starts rotating, and the magnetophoretic circuits transport the particles into the storage sites. An example of a particle trajectory in a silicon-based magnetomicrofluidic chip is shown in Fig. 7.15**a**. In this figure, the black, red, and green dashed lines stand for the particle trajectories in phases 1, 2, and 3, respectively. As shown by the green dashed line in Fig. 7.15**a**, in phase 3, after the particle moving with the magnetic forces turns into the microchannel ending to the microchamber, it is possible to help the particle transport into the microchamber by intruding a forward fluid flow. It is not necessary to apply this fluid flow, and pure magnetic forces are sufficient to transport the cells; however, the fluid flow increases the particle speed and decreases the particle assembly time.

A small section of a magnetomicrofluidic chip in which single cells are success-fully loaded in the storage sites is illustrated in Fig. 7.15**b** [10]. Based on the reported experimental results, the efficiency of the magnetomicrofluidic chip in loading single particles with sizes of ~ 10 μm or less (e.g., Human CD4+ T cells) is more than 60%. These results are plotted in Fig. 7.15**c**. The chips with 96 rows of 40 compartments in series are loaded with magnetic beads with diameters in the ranges of 5–5.9 μm, 10–13.9 μm, and 14–17.9 μm, which in Fig. 7.15**c** are called 5 μm, 10 μm, and 15 μm, respectively. For a channel height of ~ 27 μm and particles with a mean diameter of 15 μm, the majority of the storage sites are reported to remain empty. This low efficiency is due to the effect of the unwanted magnetic film, as explained above. But these particles have been loaded into a chip with channel heights of ~60 μm with an efficiency of ~ 60%, which again agrees with the rule mentioned for the appropriate particle loading.

7.4 Biocompatibility

One of the main goals of magnetomicrofluidic chips is to form single-cell arrays and then study their dynamic behavior on the chip. Hence, biocompatibility tests are an important need. The cells need to be kept alive on the chip for multiple days by providing the right conditions, including a suitable surface for the adherent cells to adhere to, the right temperature, and an appropriate pH.

Cell manipulation in any in vitro platform may affect cell behavior and viability. To evaluate how cell manipulation in the magnetomicrofluidic chips affects the cells, biostudies are conducted. In these studies, the cell behavior, including cell adhesion, growth, and drug response, under the shear stress, the magnetic nanoparticle labeling, and the microenvironment effects are evaluated.

Fig. 7.15 Loading the magnetomicrofluidic chip with particles. (**a**) The magnetic particle trajectory in a silicon-based magnetomicrofluidic chip is shown. A small section of a large array is illustrated. The particle trajectories in three phases are depicted. In phase 1, the forward fluid flow moves the particle along the black dotted line. In phase 2, the reverse fluid flow moves the trapped particle from the trap site to the magnetophoretic circuits along the red dotted line. In phase 3, the magnetophoretic circuits move the magnetic particles into the storage site, along the green dotted line. In this example, after placing the magnetic particle on its right path to the storage site with the magnetic forces, a forward fluid flow helps the magnetic forces to move the particles faster. (**b**) The chip is loaded with cells. (**c**) The chip loading efficiency is plotted. The black, gray, and blue bars depict the percentage of compartments loaded with a single particle, no particle, and a doublet (or more). © [Year] IEEE. Reprinted, with permission, from [10]

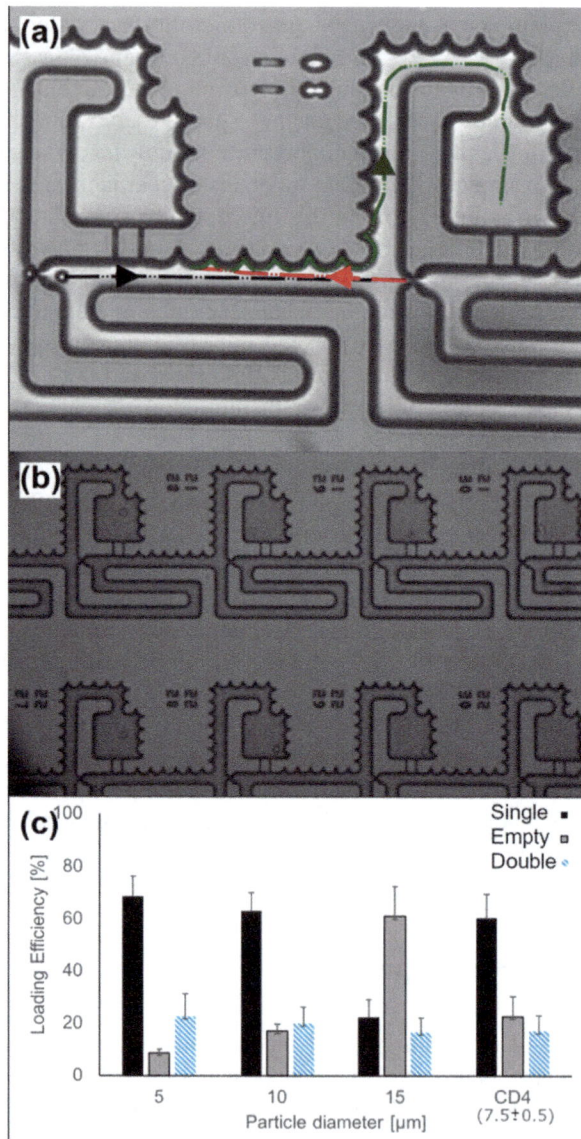

7.4.1 Shear Stress Effects

The endothelial cells located in the inner layer of the vessel walls sometimes experience relatively strong shear stresses. Also, smooth muscle cells that are usually in normal conditions protected with the endothelium, in endothelial injuries, sometimes experience shear stresses. However, the interstitial fluid moves slower than blood.

For instance, in the rabbit ear window model, this speed is reported to be less than 2 μm/s with an average of 0.6 μm/s [11]. Also, in anesthetized mice, it is reported to be in the range of 0.1–0.5 μm/s [12]. In disease conditions, these velocities may reach up to 10 μm/s [13, 14]. Thus, the reported shear stresses for various cells in different systems and conditions are different.

The cellular activities in the blood circulatory system and lymphatic capillaries are regulated in part by the mechanical stimulation provided by the fluid flow. The received mechanical stimuli are transduced into biochemical responses and modulate various parameters such as cell morphology, cell-extracellular matrix (ECM) adhesions, gene and protein expression levels, and the protein secretion profile [15–17]. For example, morphology, proliferation, and viability of the endothelial cells are regulated by the shear stress they experience. Hence, the effect of shear stress on various cell types is required to be considered in the in vitro models too. The shear stress is calculated using Eq. (7.2).

$$\nu = -\left(\frac{12Q\eta}{h^2 w}\right),\tag{7.2}$$

where Q, μ, h, and w are the volumetric flow rate, fluid viscosity, channel height, and channel width, respectively. Usually, in the literature, shear stress is denoted by τ; however, since in this text it represents the time constant, ν is used instead. Hence, this expression confirms what was mentioned before. In the magnetomicrofluidic chips, the shear stress in the trap site, which is a narrow channel, is higher than the one in wider microchannels.

In order to ensure that the cells behave normally on the chip, the shear stress applied to them in magnetomicrofluidic channels needs to be considered. In a study, the proliferation of the PC9 cells in the storage sites is investigated and compared with cells kept in the trap sites on a magnetomicrofluidic chip, while two different fluid flows are applied [18]. The results indicate better cell proliferation for the cells kept in the storage sites with lower shear stresses compared to the cells in trap sites. Similar results have also been reported by other researchers [19, 20]. In the study on the magnetophoretic circuits [18], the cell which is captured in a trap site with a high flow rate does not adhere to the chip and is dead. The cell trapped in a trap site with a lower flow rate is shown to adhere to the chip. Cells in storage sites in this study are shown to experience appropriate flow rates that proliferated. The number of cells in the storage sites increases over time. The number of cells at various flow rates is shown to be very close, which means the storage site is a good place for the cells to grow in a wide input flow rate range. But, the number of cells in the trap sites at high flow rates drops over time, which means they die.

7.4.2 Magnetic Labeling Effects

To move the cells in the magnetomicrofluidic chips with magnetic forces, they need to be magnetically labeled. Cell magnetic labeling is done by bonding magnetic nanoparticles (e.g., superparamagnetic iron oxide nanoparticles) to the cell. This process is done by ligand-receptor pairing [21] on the cell surface or by internalizing the magnetic nanoparticles into the cells [20, 22]. In various studies, the biocompatibility and cytotoxicity levels of the magnetic nanoparticles are evaluated [23, 24]. In cells with a high dose of magnetic nanoparticles, reactive oxygen species (ROS) generation is reported [25, 26], which is lowered with particle treatment [27]. I did not find any report about the adverse effects of the nanoparticles bonded to the cell surface. Some researchers have reported the biocompatibility of the nanoparticles [28].

Studies have shown that cell magnetization does not affect their growth rate and drug screening results [18]. The results indicate that no major difference between the parental cell line and its labeled counterparts is seen. These tests are performed on non-adherent MOLM-13 acute myeloid leukemia (AML) cells. The cells show similar behavior before and after being magnetically labeled. In these experiments, first, the cells are cultured and they are counted after 24 and 48 h. The cell numbers at these time points in the two cell groups (*i.e.*, labeled and unlabeled cells) are close to each other. Then, the cells are treated with quizartinib, which is a potent inhibitor of the FLT3 receptor tyrosine kinase. The cell viability at different drug doses is investigated, which for the two cell groups is similar.

7.4.3 Microenvironment Effects

Till now, we learned that by adjusting the shear stress in the microfluidic channels and by controlling the magnetic nanoparticle level for labeling the cells, their effect on the cells can be controlled. Also, the right temperature for the cells as mentioned before is typically 37 °C. Additionally, a CO_2 gas concentration of 5% typically provides the right pH for them. Temperature and gas sensors are needed to check these two parameters in the magnetomicrofluidic chips. Now, the microenvironment effect is an important parameter to make the cells "happy." By providing an environment similar to what they experience inside the body, we can give them the opportunity to survive on chips and behave normally. The extracellular matrix (ECM) is a network of proteins and other molecules surrounding the cells in the body. It is required to use a good combination of substrate and covering protein to grow the cells on the chip.

PC9 cells are cultured on PDMS, SU8, silicon dioxide (SiO_2), and silicon substrates, which are widely used in microfluidic chambers (e.g., in magnetomicrofluidic chips) [18]. PC9 is a well characterized and widely used epidermal growth factor receptor (EGFR)-mutant lung cancer cell line [22–24]. This is an adherent cell line that grows even from single cells. These two characteristics make them good candidates for initial studies. Since they are adherent, they do not move after being cultured, so they can be counted easily. Also, the ultimate goal of the magnetomicrofluidic chips is to run single-cell studies. Hence, it is important to grow the cells from single cells. In these experiments, the cells have been supplied with Gibco RPMI-1640 medium (Life Technologies) with 10% Fetal Bovine Serum (FBS). To run the tests, the substrates are placed in petri dishes, and then the cells with cell culture media are added. The experiments are repeated several times for each substrate, and different spots on each chip are investigated. Based on the morphologies of the resulting cells after 48 h of incubation, many cells die on the SU8 and PDMS substrates, which means they have not experienced the proper conditions. However, the cells adhere and grow on the silicon and silicon dioxide substrates, which makes them good candidates for the magnetomicrofluidic chips. Please note that these experiences are performed at the population level.

To see whether the SU8 and PDMS substrates can be modified to become a good place for maintaining the cells, in another set of experiments the poly-lysine-treated substrates have been tested. The results are illustrated in Fig. 7.16 (two sample images for each substrate are provided). The adhesion/growth behavior of the cells on substrates before and after treatment is very similar. The only exception is for the cells grown on SU8 substrate, which are much better after poly-lysine treatment. Thus, in addition to the silicon and silicon dioxide substrates, poly-lysine-treated SU8 substrates can also be used for growing these cells.

As mentioned above, the tests are performed at the cell population level. But the story at the single-cell level, which is what we are interested in, can be different. Hence, the experiments have been repeated at the single-cell level on silicon and silicon dioxide substrates. However, in these experiments, the cells did not adhere to the substrate. The same experiments have been repeated with the poly-lysine-treated substrates, with the same results.

Thus, when the cells are single, they behave differently compared to them at the population level. They need to "sense" other cells, which is typically done via the cellular signaling factors found in the surrounding environment. In a culture dish with a large number of cells growing in it, these factors are available in the surrounding cell culture media. So, that media may help cells grow at the single-cell level too. Hence, the single-cell culture experiment on the mentioned substrates with "conditioned media" (*i.e.*, the cell culture media taken from a dish in which the cells have been grown) has been repeated, and good results are achieved.

The results of culturing experiments at the single-cell level in conditioned media are shown in Fig. 7.17. The conditioned media helps the cells behave normally and adhere to the silicon wafer (See Fig. 7.17**a**). Similar results are reported to be seen on silicon dioxide substrates. In Fig. 7.17**b**, a microscopy image of the cells in a similar experiment on silicon wafers etched with RIE is shown. In this experiment, the substrate is even closer to the ones in the magnetomicrofluidic chips, in which the microchannels are etched into the silicon wafer. Relatively good results are achieved; however, the cells did not adhere as well as the ones on non-etched silicon wafers (compare the cells in Fig. 7.17**a** and **b**).

Fig. 7.16 The PC9 cell
populations are cultured on
the poly-lysine-treated
substrates. The PC9 cells are
grown on (**a, b**) PDMS, (**c,
d**) SU8, (**e, f**) SiO2, and (**g,
h**) silicon (treated) substrates
and incubated for 48 h

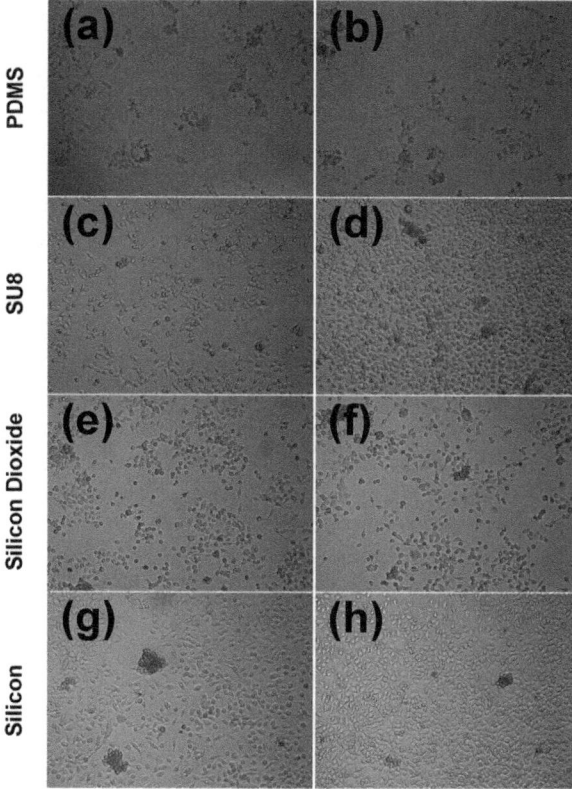

To answer this problem and provide a more appropriate microenvironment for the
cells to better adhere to the substrates, the etched silicon and silicon dioxide wafers
are coated with fibronectin, collagen, and gelatin. In these tests, conditioned media
are provided to the cells every 48 h. Also, to better mimic the magnetomicrofluidic
chips, in silicon dioxide substrates, a thin film of permalloy was included underneath
the silicon dioxide layer. Also, both silicon and silicon dioxide chips were etched.
In these experiments, PC9 and 3T3 cells are tested. 3T3 is a widely used mouse
fibroblast cell line [29, 30]. These cells adhere to the culture dish and can grow from
single cells. Thus, they are good candidates for single-cell studies.

Promising results for the PC9 cells cultured on collagen and gelatin-treated
substrates are observed. The ratio of the adhered cells to the ones which did not
succeed in adhering to the chip at various time points indicates that the cells mostly
adhere to the substrates on day 2. Based on these studies, 3T3 cells mostly adhere
faster than the PC9 cells to the silicon substrates; however, they do not adhere well
to the silicon dioxide substrates.

Cell growth is another important factor to be considered. Studying the average
cell density at different time points indicates that most of the cells grow starting
from day 2. But, as expected, 3T3 cells, which do not adhere well to silicon dioxide

Fig. 7.17 PC9 cells are grown on bare silicon substrates at the single-cell level in conditioned media. The cell morphology is shown on (**a**) a bare silicon substrate and (**b**) an etched silicon wafer. The cells are incubated for 48 h

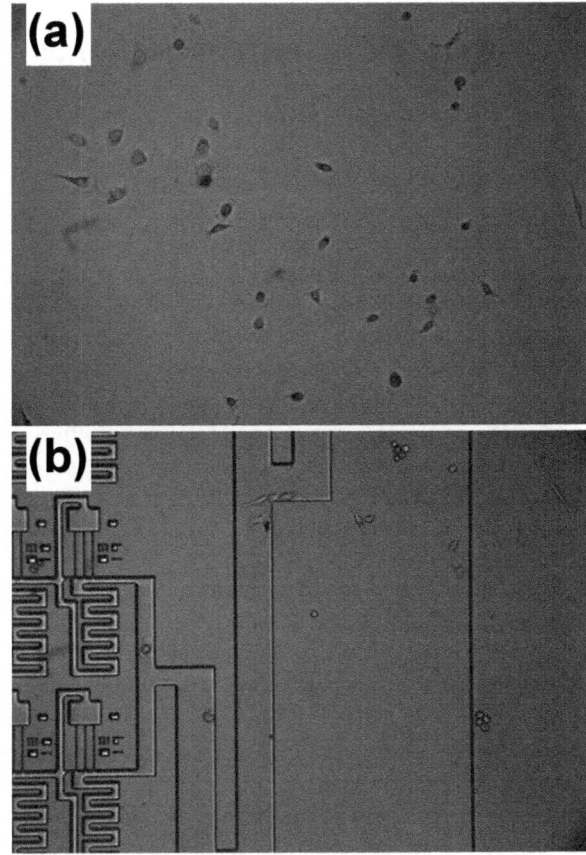

substrates, do not grow, even after 96 h. To calculate the cell density, the cells in a unit surface area are counted, which can be done in two different methods. First, it is possible to simply count the cells under a microscope. The advantage of this method is that the same cells can be counted at multiple time points, without being killed or manipulated. In the second method, the cells are trypsinized, stained, and then counted. In this method, cell counting is easier; however, for evaluating the cell numbers at different time points, the experiments need to be repeated, either with different cells or with the counted cells, none of which are preferred.

Sample microscope images of PC9 and 3T3 cells which are cultured on various substrates at different time points are shown in Figs. 7.18 and 7.19, respectively. These experiments have been started from single cells. In these figures, most cells have adhered to the substrates after day 2 (*i.e.*, 48 h). As mentioned, the PC9 cells grow well on both silicon and silicon dioxide substrates; however, the 3T3 cells do not adhere or grow well on silicon dioxide substrates. The last row in Fig. 7.18 shows images of a control experiment in a petri dish. Comparing the column 96 h of different rows with that of the petri dish experiment shows similar behavior and morphologies are achieved. But in the case of 3C3, it is not like that. The petri dish is full of cells after 96 h (See the last row in Fig. 7.19). The results for the cells grown on silicon wafers are almost similar to the ones in petri dishes. But the cells are too few on silicon dioxide wafers (See the first two rows in Fig. 7.19). Hence, not all substrates are suitable for all cells.

Now that we know how the cells behave on the chosen substrates, let's see how they behave inside magnetomicrofluidic chips (closed chips). The microfluidic chips are treated by injecting the solution of interest into the microchannels. In the case of collagen, since it is prepared in acetic acid, the cells cannot be introduced to the chip right after treatment. But gelatin, prepared in deionized water, can be used for treatment, after which a short rinse with conditioned media helps.

In experiments with the gelatin-treated PDMS-based microfluidic chips, some cells adhere to the chips, but some don't. The cells adhere to the silicon/glass-based magnetomicrofluidic chip after 48 h. However, they do not survive more than 72 h. Treating the chips and providing the right CO_2 gas concentration in the media helps in increasing the cell viability. If syringe pumps are not available, a 1000-μL pipette tip carrying conditioned cell culture media can simply be inserted into the inlet port. This setup produces a very slow gravity-driven fluid flow inside the microchannels, and the chip can be kept in an incubator.

In conclusion, by integrating magnetophoretic circuit elements, various circuits for different applications can be designed to transport magnetic particles in parallel to specific spots on the chip. The particles are synched with the external rotating magnetic field, which provides good control on a large number of them. The individual particles can be switched at specific points with scalability and control, similar to the ones found in electronic chips. By coupling the magnetophoretic circuits with microfluidic channels, rapid precise particle transport is achieved. It provides the opportunity to assemble particles, cells, and potentially multi-component patterns of cells into high-density arrays.

Since PDMS-based magnetomicrofluidic chips have problems in handling some solvents, the novel magnetomicrofluidic chips are fabricated based on silicon and glass. But because of the problems in forming magnetic thin films inside the microchannels, a new chip design and a novel microfabrication protocol are used. In this method, the magnetophoretic circuits are shifted into the microfluidic walls. The microchannels are etched in the silicon wafer, and the anodic bonding technique is used to seal it with a glass slide.

The capability of the magnetomicrofluidic chips in maintaining the cells alive for a few days is another important study. Thus, the cell behavior in the microfluidic chips has been studied to provide the right conditions for the cells. Microfluidic chips based on silicon and glass coated with collagen are shown to be a good choice for culturing cells such as the PC9 cell line. For any cell type, an appropriate environment is needed. Moreover, to grow the cells from a single-cell level, conditioned media helps.

In order to complete this system, suitable pumps to provide enough conditioned media to the cells at the right flow rate are needed. The applied shear stress on the

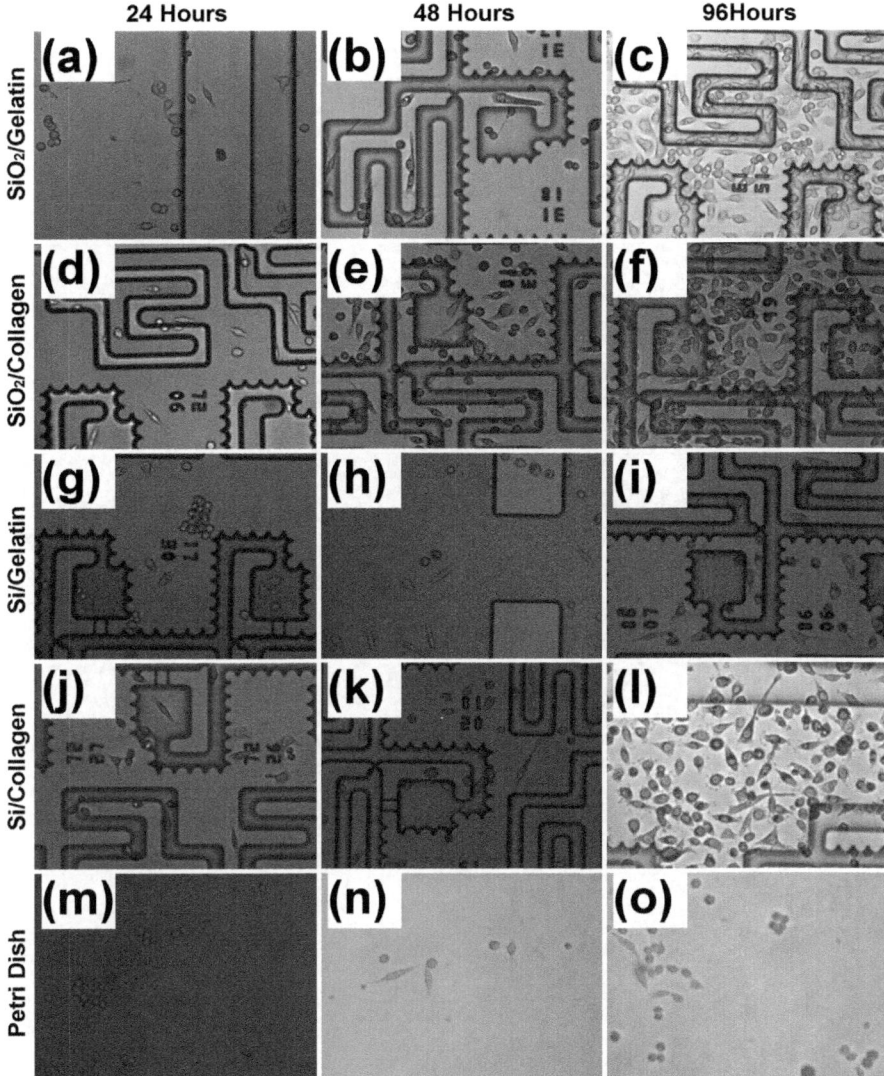

Fig. 7.18 Sample microscopy images of PC9 cells cultured at the single-cell level on treated substrates in conditioned media are illustrated. The cells are grown on (**a, b, c**) silicon dioxide substrates treated with gelatin, (**d, e, f**) silicon dioxide substrates treated with collagen, (**g, h, i**) silicon substrates treated with gelatin, (**j, k, l**) silicon substrates treated with collagen, and (**m, n, o**) culture dishes (control). The left, middle, and right columns show sample pictures after 24, 48, and 96 h, respectively

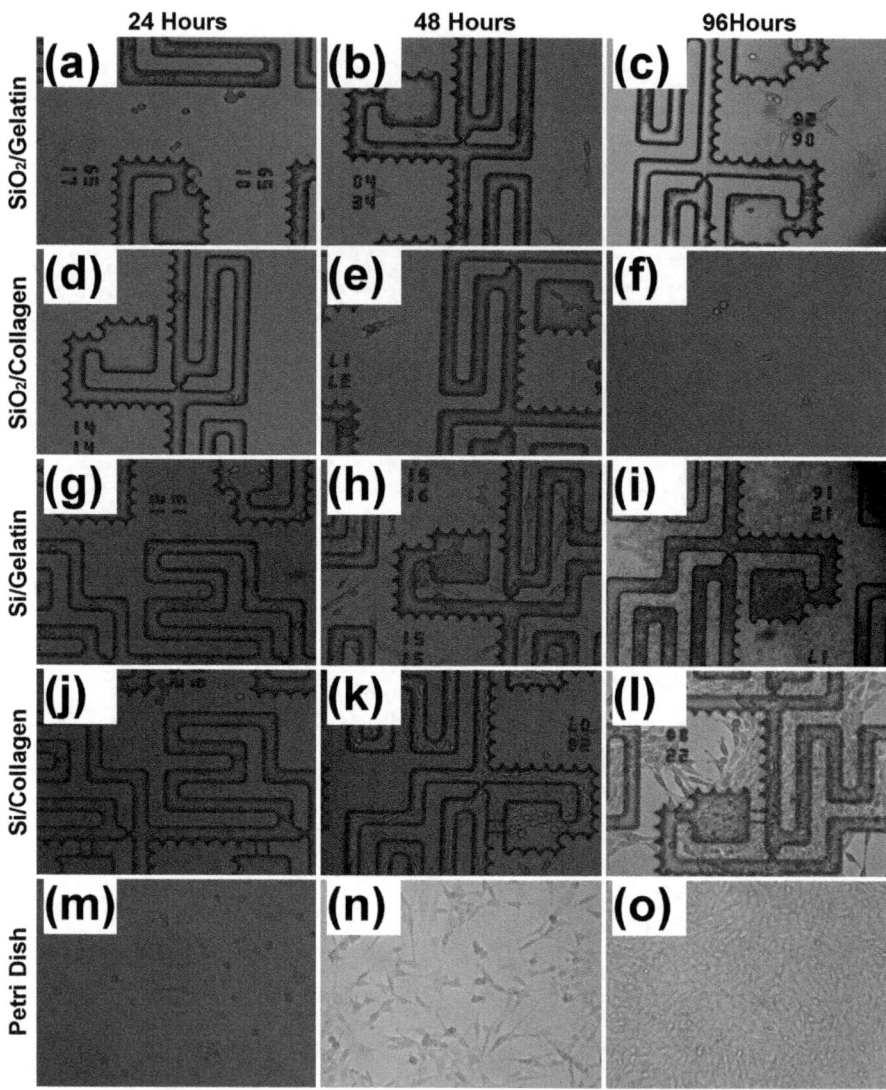

Fig. 7.19 Sample microscopy images of 3T3 cells cultured at the single-cell level on treated substrates in conditioned media are illustrated. The cells are grown on (**a, b, c**) silicon dioxide substrates treated with gelatin, (**d, e, f**) silicon dioxide substrates treated with collagen, (**g, h, i**) silicon substrates treated with gelatin, (j, k, l) silicon substrates treated with collagen, and (**m, n, o**) culture dishes (control). The left, middle, and right columns show sample pictures after 24, 48, and 96 h, respectively

cells is an important parameter to be considered to maintain a high level of cell viability. The temperature and pH have to be monitored.

7.5 Applications

Magnetophoetic circuits can be used in various applications. Developing a single-cell analysis based on the magnetophoretic circuits is one of the main goals. Towards that goal, many pilot studies have already been run, and promising results are achieved. The chips are loaded with cells where the cells are kept alive for multiple days [10]. The cells start growing from single cells (See Fig. 7.20a, b). Figure 7.20b illustrates a cell clone proliferated from a single cell, which then can be used for further biological studies.

As a drug screening test, the effect of quizartinib and ponatinib on MOLM-13 cells is shown in Fig. 7.20c, d [10]. In these figures, the drug effects on the cells in the magnetomicrofluidic chips are compared with the drug sensitivity tests in dishes.

Also, some pilot gene expression analysis has been done to study the CYP1A1 expression induction by PCB 126, which is a potent ligand for aryl hydrocarbon receptor (AhR). The results for HepG2, Hela, and HaCaT cell lines are illustrated in Fig. 7.21. The results for the cells cultured and treated in the magnetomicrofluidic chip agree with the ones in culture dishes. Treatment with 3 μM PCB 126 for 6 h leads to significant CYP1A1 mRNA expression induction. Less induction of CYP1A1 mRNA expression in treated HeLa cells is reported. These results also agree with the bulk level findings [31, 32].

The proposed magnetomicrofluidic chips can be used in other biosensing applications too. For example, its capability in single-molecule detection is shown. This tool is shown to detect herpes simplex virus type 1 (HSV-1, or oral herpes) without any need for amplification. HSV is a DNA virus of the Herpesviridae family that affects more than 60% of the global population. In this method, the single molecules are detected based on the reaction of the analyte of interest (e.g., HSV UL27 gene) between two magnetic beads. The DNA fragments are labeled with biotin and digoxigenin oligonucleotide probes to bind between streptavidin and anti-digoxigenin antibody-labeled magnetic beads and form a magnetic bead pair or cluster. Thus, forming the bead pairs or clusters indicated the presence of the analyte of interest. This chip is able to detect multiple analytes simultaneously. Hence, magnetomicrofluidic chips are also used to detect biotinylated bovine serum albumin (BSA). For detecting BSA, binding between streptavidin-coated magnetic beads is used. In Fig. 7.22, this application of the magnetomicrofluidic chips is presented.

In a pure horizontal field, as discussed before, the magnetic particles may form pairs and cluster, even in the absence of the analyte of interest. Hence, revealing the real number of analytes of interest becomes problematic. Thus, in this design, drop-shape magnetophoretic circuits operating in a tri-axial magnetic field are used. The vertical magnetic field component in this system biases the beads to repel each other and do not form unwanted particle pairs. This repulsion force is shown in

Fig. 7.20 Single-cell proliferation and drug sensitivity tests are shown. (**a**) A single cell in a magnetomicrofluidic chip is shown. (**b**) The cells grow after four days. Drug screening test results for (**c**) quizartinib and (**d**) ponatinib are shown, and compared with the experiments in the petri dishes. Inset shows samples of living and dead single cells in the chip. © [Year] IEEE. Reprinted, with permission, from [10]

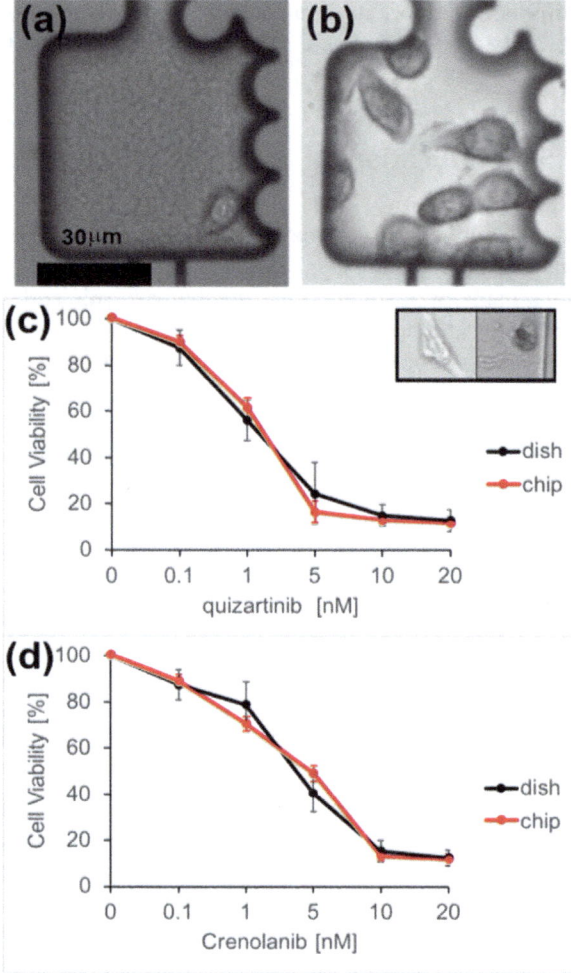

Fig. 7.22**a**. In Fig. 7.22**b–d**, the possible links between the beads to form bead pairs are shown. As illustrated in Fig. 7.22, the particles move to the detection spots that are microchambers with different analytes and various concentrations. It is shown that the number of bead pairs is proportional to the analyte concentration in each microchamber on the magnetomicrofluidic chip. As presented in Fig. 7.22**f, h**, the experimental results based on the magnetomicrofluidic chip agree with the results of flow cytometry. It is also shown that, as expected, 2D magnetic fields are not suitable for this system.

Another application of magnetomicrofluidic chips is magnetometry. The magnetic susceptibility of single magnetic particles, including the magnetically labeled cells, is an important parameter to be measured. This parameter indicates their level of response when exposed to an external magnetic field. The available methods for

Fig. 7.21 Dose–response evaluation of CYP1A1 mRNA expression using the magnetomicrofluidic chip is shown. Relative CYP1A1 mRNA levels in (**a**) HepG2, (**b**) HaCaT, and (**c**) HeLa cell lines cultures in the magnetomicrofluidic chips (black) are compared with the ones cultured in dishes (gray) (normalized to the RPLP0 mRNA levels). © [Year] IEEE. Reprinted, with permission, from [10]

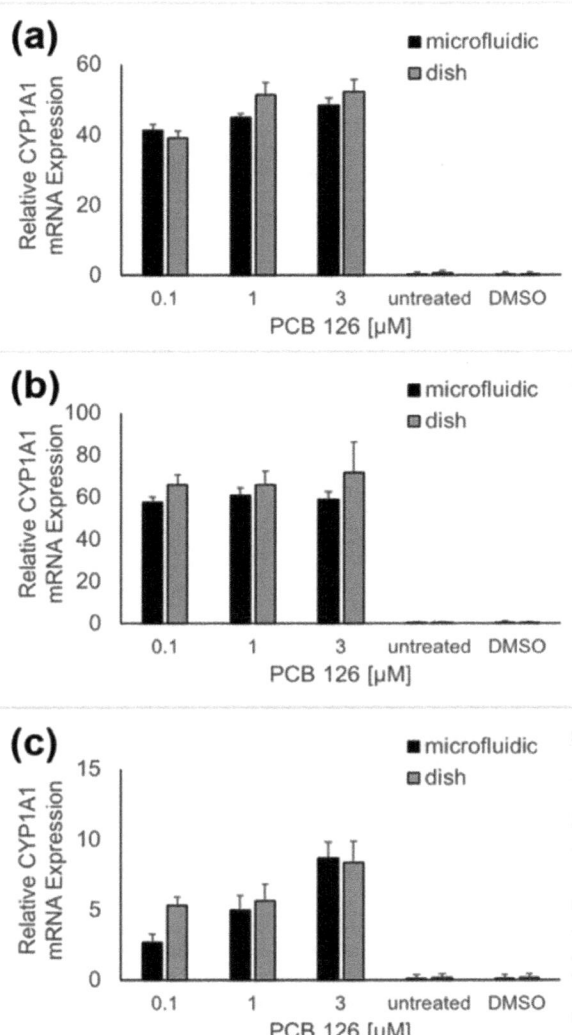

measuring this parameter are based on the superconducting quantum interference device (SQUID) [34, 35], micro-Hall sensors [36, 37], vibrating sample magnetometry (VSM) [38], and magnetophoretic mobility-based measurement [39]. Most of the measurements are average-based methods. It is important to note that the bulk measurement is challenging since (i) more material is used for measurement, which becomes challenging when the available material is limited, and (ii) single-particle resolution is not offered.

In the previous chapters, it was mentioned that the particle trajectory around the magnetic disks is a function of their magnetic content. Hence, the phase transition

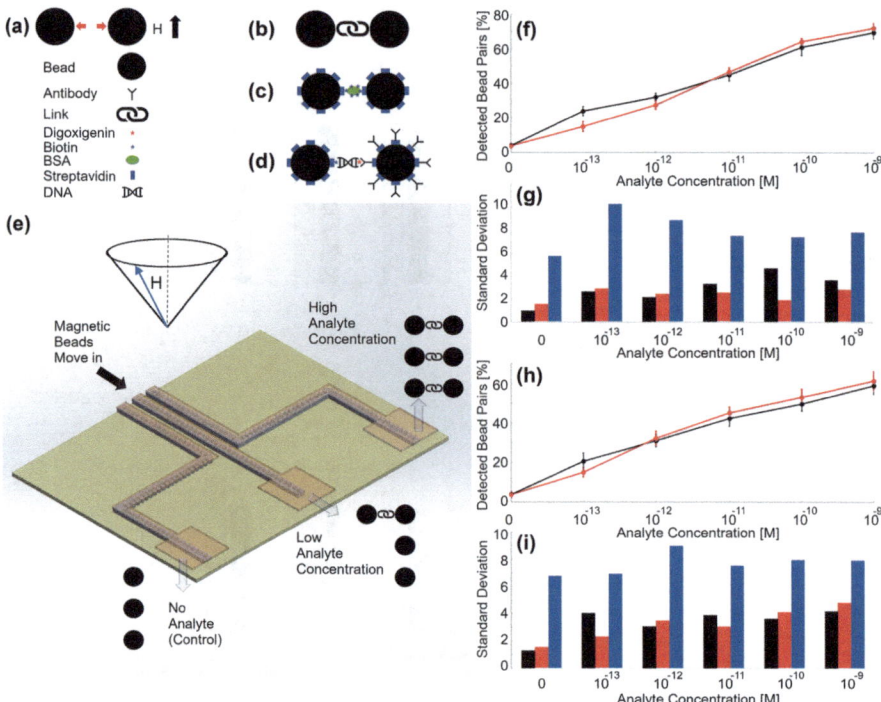

Fig. 7.22 Schematic of the biomolecule detection system based on the magnetomicrofluidic chips is shown. (**a**) The repulsion force between two particles in a vertical bias field is presented. (**b**) The two particles with a link between them form a particle pair. (**c**) A protein forms the link between the two particles. (**d**) A DNA fragment provides the link between the two particles. (**e**) A schematic of the magnetomicrofluidic chip for moving the detecting magnetic beads to the analyte-containing microchambers is shown. H stands for the applied tri-axial magnetic field. (**f**) The bead pair percentage versus the BSA concentrations, based on the magnetophoretic circuits (black) and flow cytometry (red), are plotted. (**g**) The standard deviation of the percentage of the detected particle pairs in BSA analysis based on flow cytometry (black), magnetomicrofluidic circuits in a tri-axial magnetic field (red), and magnetomicrofluidic circuits operating in a 2D magnetic field (blue) are plotted. (**h**) The bead pairs' percentage versus the HSV DNA concentrations, based on the magnetophoretic circuits (red) and flow cytometry (black), are presented. (**i**) The standard deviation of the percentage of the detected particle pairs in HSV DNA analysis based on flow cytometry (black), magnetophoretic circuits operating in a 3D magnetic field (red), and magnetophoretic circuits operating in a 2D magnetic field (blue) are plotted. The figure is reprinted from [33] under a Creative Commons Attribution 4.0 International License. http://creativecommons.org/licenses/by/4.0/

above the critical frequency is used to determine their magnetic susceptibility [40]. This phenomenon is the basic idea for magnetomicrofluidic magnetometry. In this method, the magnetic susceptibility of numerous individual particles (at the single-particle resolution) simultaneously in parallel is measured.

The particle magnetic susceptibility measurement in this method is done in two steps. First, the particle motion phase diagram is obtained. Then, the critical frequency is measured experimentally and matched with this phase diagram to determine the magnetic susceptibility of the particle. The first step is done based on particle trajectory simulation results. These simulations show results for particles of various sizes and magnetic susceptibilities. The simulations start at low frequencies, and then this parameter is increased to find the critical frequencies. These simulations are performed once, and then the obtained phase diagrams are used in future measurements.

The second step is performed by recording the particle trajectories. In these experiments, the external magnetic field frequency is increased until entering the phase-slipping regime. The particle trajectory analysis is done using image processing software. This software, after finding the micromagnets and the circulating beads, detects the frequency at which the distance of the particle from the center of the magnet is increased, and a slip-out trajectory is obtained. This frequency is booked as the critical frequency, and based on the phase diagram obtained in step 1, the magnetic susceptibility of the particles is evaluated. See the detection process block diagram in Fig. 7.23.

Two examples of the critical frequency phase diagrams for particles with diameters of 2.8 μm and 8.4 μm are presented in Fig. 7.24, which are obtained based on particle trajectory simulation analysis. The curves in these plots stand for the border between the region below them (*i.e.*, the phase-locked regime) and above them (*i.e.*, the phase-slipping regime).

Now, the plots in Fig. 7.25 are used to find the magnetic susceptibility of the particles, using the protocol presented in Fig. 7.23 for the experimental step. Figure 7.25 shows the output scatter plot for the magnetic susceptibilities of numerous particles of different sizes. In this experiment, in addition to the commercial magnetic beads, magnetically labeled CD4+ T cells (red diamonds in Fig. 7.25) are also characterized. The results are evaluated with SQUID measurements and shown in Table 7.1.

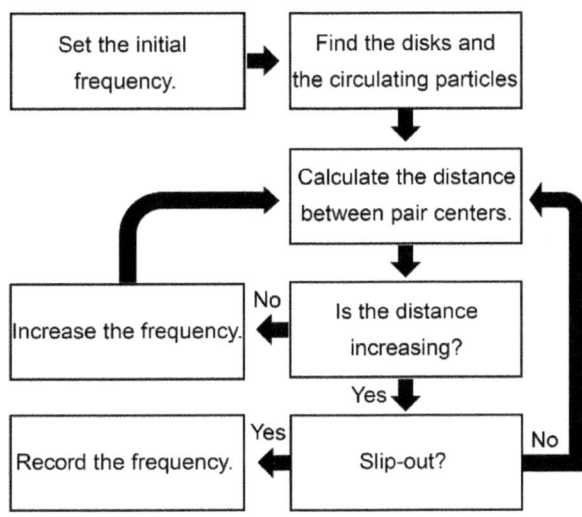

Fig. 7.23 The block diagram of the phase change detection steps in the software developed in [40]. The figure is reprinted from [40] with permission from the Royal Society of Chemistry

Fig. 7.24 The bead motion phase diagrams based on particle trajectory simulations are illustrated. The particle size is (**a**) 2.8 μm and (**b**) 8.4 μm. The curves stand for the border between the region below them (*i.e.*, the phase-locked regime) and above them (*i.e.*, the phase-slipping regime). The figure is reprinted from [40] with permission from the Royal Society of Chemistry

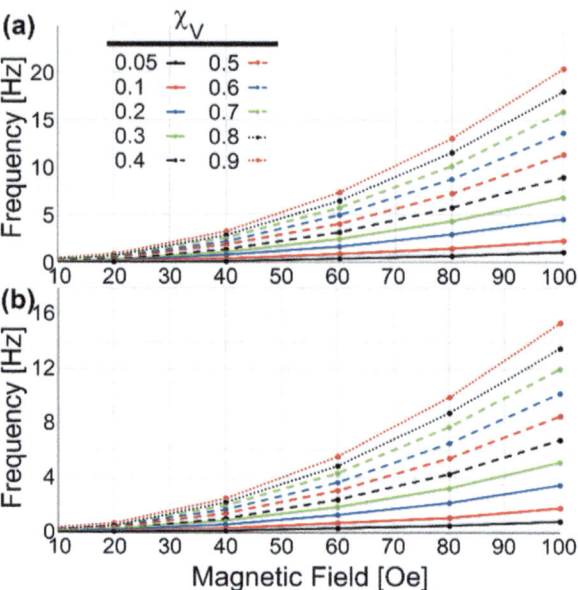

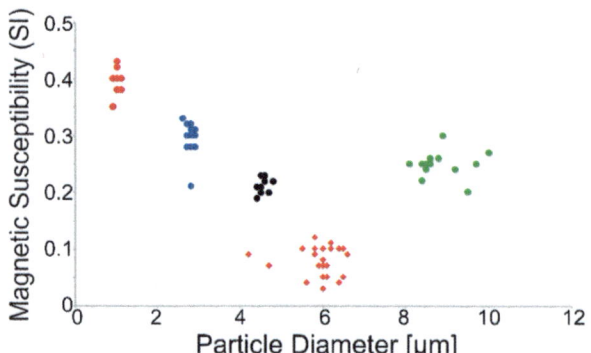

Fig. 7.25 The scatter plot shows the magnetic susceptibility of the particles with different diameters, obtained using the magnetophoretic-based device. Each circle or diamond stands for a single particle in a single test. The red, blue, black, and green circles represent the MyOne™, M-280, M-450, and FCM-8056-2 magnetic particles, respectively. The red diamonds stand for the CD4+ magnetically labeled cells. The figure is reprinted from [40] with permission from the Royal Society of Chemistry

As mentioned before, magnetometers have a wide range of applications in various fields. An example in the field of bioengineering is to identify the magnetization of the cells which uptake magnetic nanoparticles over time. Figure 7.26a shows the magnetic susceptibility measurement results of the magnetophoretic-based device for two different cell types incubated with magnetic nanoparticles for up to 120 min. Then

Table 7.1 Magnetic susceptibilities of particles based on the magnetophoretic chip and SQUID

Particle	Susceptibility	Measurement Method	Reference
MyOne™	0.34	SQUID	[38]
	0.4	SQUID	[40]
	0.4	Magnetophoretic	[40]
M-280	0.15	SQUID	[38]
	0.7	SQUID	[37]
	0.3	SQUID	[40]
	0.3	Magnetophoretic	[40]
M-450	0.25	SQUID	[41]
	0.21	SQUID	[42]
	0.2	SQUID	[40]
	0.21	Magnetophoretic	[40]
FCM-8056-2	0.3	SQUID	[40]
	0.25	Magnetophoretic	[40]

The table is reprinted from [40] with permission from the Royal Society of ChemistryThe table is reprinted from [40] with permission from the Royal Society of Chemistry

the magnetic susceptibility of the magnetized cells losing their magnetic nanoparticles over time is measured, the results of which are shown in Fig. 7.26**b**. These experiments provide insights into intracellular degradation of the nanoparticles at the single-cell level.

The chips also have been used in a model to identify the number of specific cells in a population. In this experiment, mixtures of cells incubated with magnetic nanoparticles for 30 min and 60 min at various ratios are prepared. Then the chip is used to identify them. The results are shown in Fig. 7.26**c**.

Although not designed initially for this application, the proposed magnetophoretic circuits can separate the particles based on their characteristics, which is an important application in the field of lab on a chip. This capability is shown in both 2D and 3D magnetic fields.

This design separates the magnetic particles based on their size and magnetic susceptibility. In Fig. 7.27**a–d**, again the energy landscapes of the magnetic track in a rotating in-plane field are shown. Also, Fig. 7.27**e–h** illustrate the corresponding trajectory of a magnetized T cell along the magnetic track. As discussed before, at higher magnetic field frequencies the particles move faster (See Fig. 7.27**i**). But at frequencies higher than the critical frequency, the movement enters the phase-slipping regime, and the particles cannot follow the external field. Thus, at these frequencies, the particle velocity drops. Since the critical frequencies depend on the size and the magnetic content of the particles, a frequency at which the particles with different characteristics move at different speeds can be found. This method leads to particle separation. Figure 7.27**j–n** depict the sequences of separating a few particles. Comparing these snapshots, we realize that the big particle remains while the others separate and move to the right. Starting with Fig. 27**j**, all the magnetic particles are on the left. Then, the small particles are transferred to the right. We

Fig. 7.26 (**a**) The magnetometer is used to measure the magnetic susceptibility of the cells that uptake magnetic nanoparticles as a function of the incubation time. The red and blue curves stand for the magnetic susceptibility of the THP-1 and HeLa cells, respectively. (**b**) The magnetometer is used to measure the magnetization of the magnetized cells losing their magnetic contents over time. The red and blue curves stand for the magnetic susceptibility of the THP-1 and HeLa cells, respectively. (**c**) The magnetometer is used to find the ratio of a target cell in a mixture. It shows the detected ratio versus the real ratio, respectively. The figure is reprinted from [40] with permission from the Royal Society of Chemistry

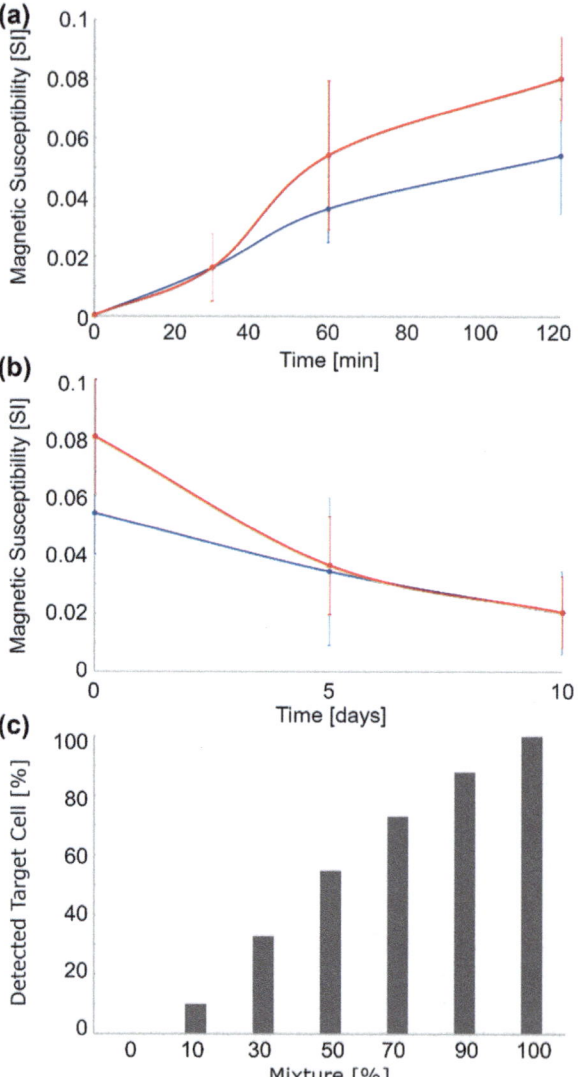

compare the two small transported particles depicted by the red and green arrows with the large particle shown by the blue arrow on the left. The red dotted line in these figures depicts a sample small particle trajectory. The separation efficiency in the frequency range of 0–10 Hz is shown in Fig. 7.27**o**.

The semiconducting gap that is used in the transistor design can also be used in separating particles based on their sizes [43]. We recall that the parameter $\beta = r_p/r_G$ was defined, where r_p and r_G are the particle radius and the magnetic track gap size, respectively. Hence, β varies for particles of different sizes. In a conducting path with a gap, a small particle cannot cross over it and remains on the initial magnetic track.

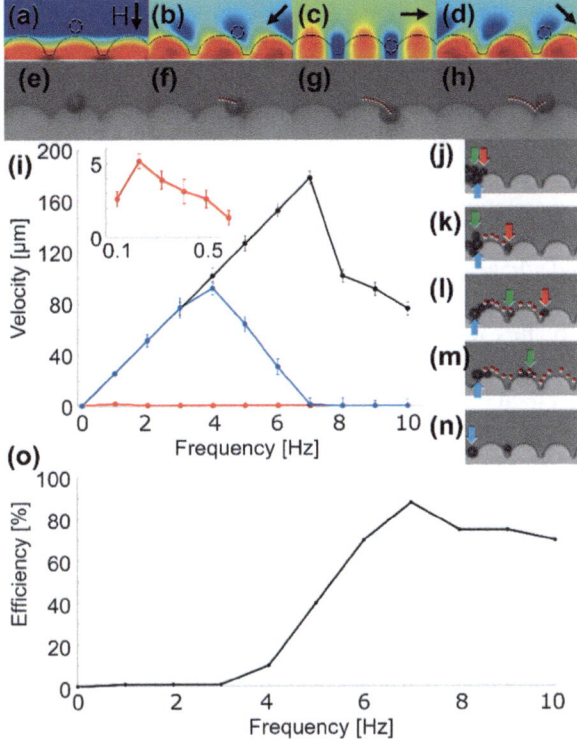

Fig. 7.27 Magnetophoretic particle separation is shown. (**a–d**) Magnetic energy landscapes at various field angles (shown by the black arrows) are shown. The black dashed circle in each panel depicts a hypothetical magnetic particle. (**e–h**) The corresponding experimental microscopy images of the T-cell trajectory along the magnetic track are shown. (**i**) The particle velocity versus frequency plot for the particles with mean diameters of 2.8 μm (black), 4.8 μm (blue), and T cells (red) are presented. The inset shows a zoomed view of the T-cell curve. (**j–n**) A few particles are on the left, and by applying a rotating in-plane magnetic field, they separate. Each panel shows a snapshot time frame of the separation. The red and green arrows show two sample small particles, while the blue arrow shows a large particle. The red dotted line depicts a sample small particle trajectory. (**o**) The particle separation efficiency versus deriving frequency is plotted. The figure is reprinted from [40] with permission from the Royal Society of Chemistry

But a large particle can cross over the gap and move to the magnetic track on the other side of it. This idea is used to separate particles based on their sizes.

To separate two groups of particles, one gap in the magnetic track is enough. But to separate more particles, more gaps of different sizes are needed. The difference between the gap sizes depends on the particle size difference. When the particle size differences are large enough (e.g., 2.8 μm and 8.4 μm particles), the gap size difference is also relatively large. But, when the particles are close in size, gaps with little size differences (e.g., in the micrometer range) are needed. However, fabricating gaps at micrometer resolution using conventional microfabrication tools is challenging. Hence, separating the particles with small size differences based on this method is problematic.

To overcome this challenge, the magnetic disk next to the gap is replaced with an ellipsoid (See Fig. 7.28) [43]. The energy distribution calculation of an ellipsoid is explained in Chap. 3. Similar to the I bar (in the TI magnetic pattern), the ellipsoids show strong magnetization when the external magnetic field is applied along their main axis. Their magnetization, when the external field is aligned with its minor axis, is weak, which can be tuned by adjusting the aspect ratio of the ellipsoid. Hence, the ellipsoid is placed adjacent to the gap such that its major axis is perpendicular to the magnetic track.

An example of this design is shown in Fig. 7.28a, where separating particles of three sizes is demonstrated [43]. The particles are sequentially separated from different gaps and then move to separate capacitors labeled I, II, and III. The largest particles with a mean diameter of 8.6 μm are assembled into capacitor I by crossing over a gap (row I in Fig. 7.28a) with an ellipsoidal magnet with an aspect ratio of 1.5. The smaller particles do not cross the first gap and move to the middle row. The particles with mean diameters of 6.7 and 3.6 μm cross over the gaps in rows II and III and move into capacitors II and III, respectively. The aspect ratio of the ellipsoidal magnets on rows II and III are 1.75 and 2, respectively. The separation efficiencies are plotted in Fig. 7.28b. The average separation efficiency of 88.86% (± 4.15) is reported.

By adding transistors to this design, forming particle pairs in the capacitors is shown too. Particle pair formation has been discussed before; however, here the two particle types enter from different ports into the capacitor (See Fig. 7.29) [43]. In this experiment, THP-1 and MCF-7 cells form cell pairs. First, in a clockwise rotating external magnetic field, the THP-1 cells (called the effector cells in Fig. 7.29) are size separated (See Fig. 7.29a). Then, an appropriate electric signal is applied to the gate of the transistor in a short time to only transfer a single particle into the capacitor (See Fig. 7.29b), after which the transistor is turned off. Then, in a counterclockwise rotating magnetic field, MCF-7 cells (called the target cells in Fig. 7.29) are transported into the capacitor from the transistor on the other side (See Fig. 7.29b). The duration and timing of the transistor gate signals define the number of particles getting into the capacitor from each side. Figure 7.29c demonstrates a table in which multiple combinations of the THP-1 cell and fluorescently labeled (green) MCF-7 cells are assembled.

The drop-shape magnetic track that operates in a tri-axial field can also separate the magnetic particles based on their size by modulating the magnetic cone angle and strength [44]. The phase diagram in Fig. 7.30 represents the percentage of magnetic particles which move in open trajectories as a function of the external magnetic field strength and cone angle. The diamonds in this plot depict the data for particles with a mean diameter of 5.6 μm. Also, the circles in this plot stand for the data for the particles with a mean diameter of 8.4 μm. The solid and dotted lines are drawn for the two particle groups to guide the eye and define the boundaries of their respective transport regimes.

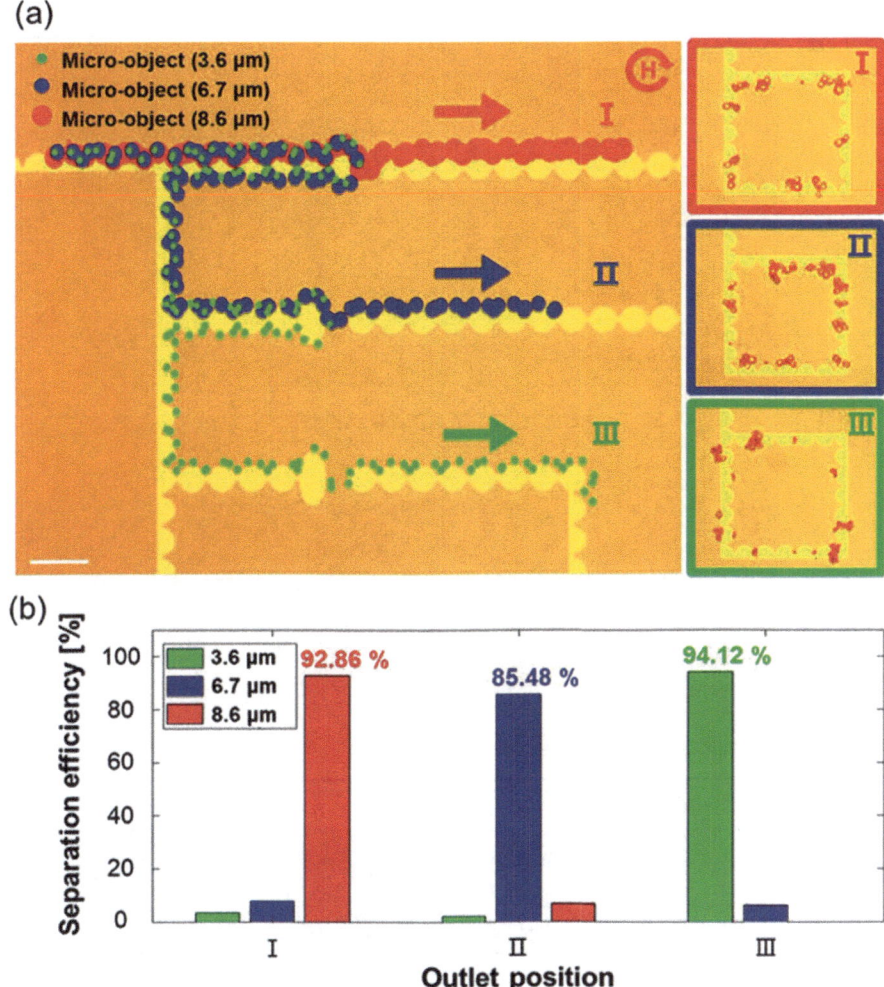

Fig. 7.28 Magnetophoretic particle separation based on the semiconducting gap is illustrated. (**a**) The trajectory of the three particle groups with sizes of 8.6 μm (red), 6.7 μm (blue), and 3.6 μm (green) are shown. They eventually move to their specific capacitors. The external field frequency is 0.68 Hz, and the field strength is 100 Oe. The aspect ratios of rows I, II, and III are 1.5, 1.75, and 2.0, respectively. (**b**) The separation efficiencies are plotted. The figure is taken with permission from [43] under the terms of the Creative Commons CC BY license. https://creativec ommons.org/licenses/

Based on this phase diagram, the combination of the cone angle of $\alpha = 37°$ and the field strength of around 50 Oe is a good choice for separating the two groups. This prediction is validated by testing a mixture of the two magnetic particle groups with the chip in the specified conditions. Figure 7.31 shows that the larger particles move in open trajectories (*i.e.*, transported to the right), whereas the smaller particles move

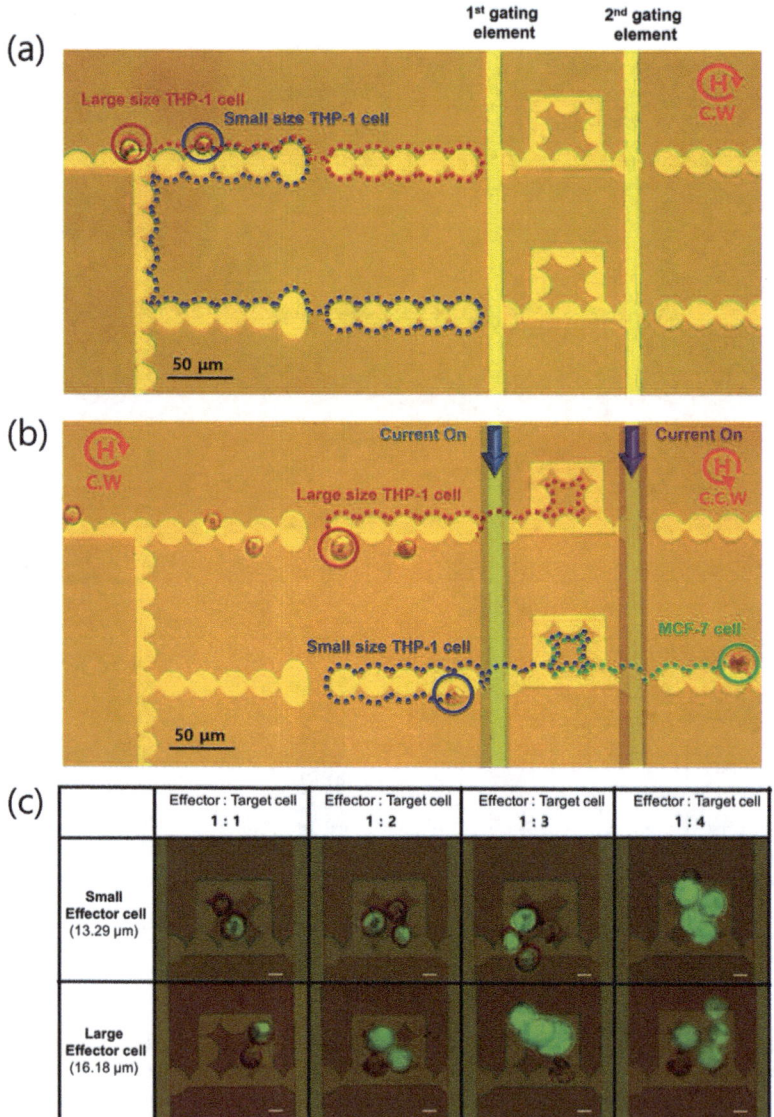

Fig. 7.29 Assembling cell pairs into the capacitors from both sides via transistors is illustrated. (**a**) THP-1 cells are size separated. The frequency is 0.18 Hz, and the magnetic field strength is 130 Oe. (**b**) The selected cell is loaded into the capacitor by signaling the gate of the transistor on the left side of the capacitor. After transporting one THP-1 cell, when the magnetic field rotates counterclockwise, the MCF-7s cell moves into the capacitor by triggering the transistor on the right side of the capacitor. (**c**) A table of the multiple cell combinations is shown. The THP-1 and MCF-7 cells are named effector cells and target cells, respectively. The figure is taken with permission from [43] under the terms of the Creative Commons CC BY license. https://creativecommons.org/licenses/

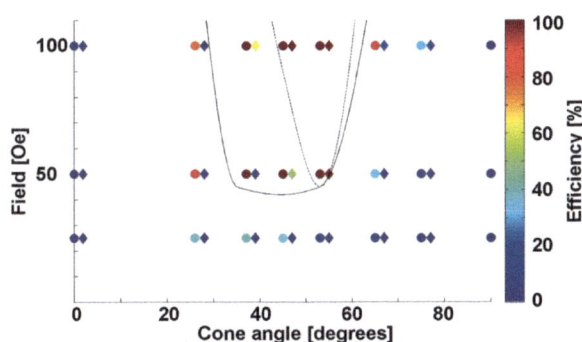

Fig. 7.30 The field versus magnetic field cone angle phase diagram on the drop-shape magnetophoretic conductor is presented for 5.6 μm diameter magnetic particles (diamonds) and 8.4 μm diameter magnetic particles (circles). The conducting and non-conducting regime boundaries for particles are drawn with a dotted line and a solid line for the small and large particles, respectively. The figure is taken with permission from [44]

in closed orbits. The blue and red dotted lines in this figure depict the trajectories of large and small particles, respectively. Hence, at the end, the small particles remain on the chip while the large particles gather on the right.

Another application of the magnetomicrofluidic circuits is in concentrating particles, which is important in the field of biosensing [45]. The biomolecules of interest are magnetically labeled with magnetic particles to be detectable in devices such as magnetoresistive sensors. Collecting the particles and carrying analytes on the sensor area is an important task in the field of lab on a chip. Magnetophoretic circuits are used for the effective collection of biomolecule carriers by forming magnetic patterns resembling a spider web network [45]. The magnetophoretic tracks based on magnetic half-disks are arranged towards the center of the circular web, where the MR sensor is integrated (See Fig. 7.32).

This magnetophoretic track arrangement results in a high-concentration area for the particles at the center. Also, they are diverged away from that point by inverting the external magnetic field rotation. The basic idea in designing this circuit relies on employing T-junction diodes. In the design shown in Fig. 7.32, in a clockwise rotating magnetic field, the diodes pass the magnetic particles to move towards the center of the web (*i.e.*, the sensor). Figure 7.32**b** shows a zoomed view of the particle trajectories along the web network junction, where the white arrows and the black dashed rectangle depict the particle trajectories and the linear track around the junction, respectively. Figure 7.32**c** shows a fluorescent image of the particles at the center of the design after concentration. In this design, applying a counterclockwise rotating magnetic field moves the particles away from the center (See Fig. 7.32**d**).

In a counterclockwise rotating magnetic field, the particles do not cross the T-junction diode and move along the curved dispersing pattern which is shown with the rectangles in Fig. 7.32**e**. Without the T-junction diodes, the particles would have gone to the outside edge of the circuit. But employing the T-junction diodes results in

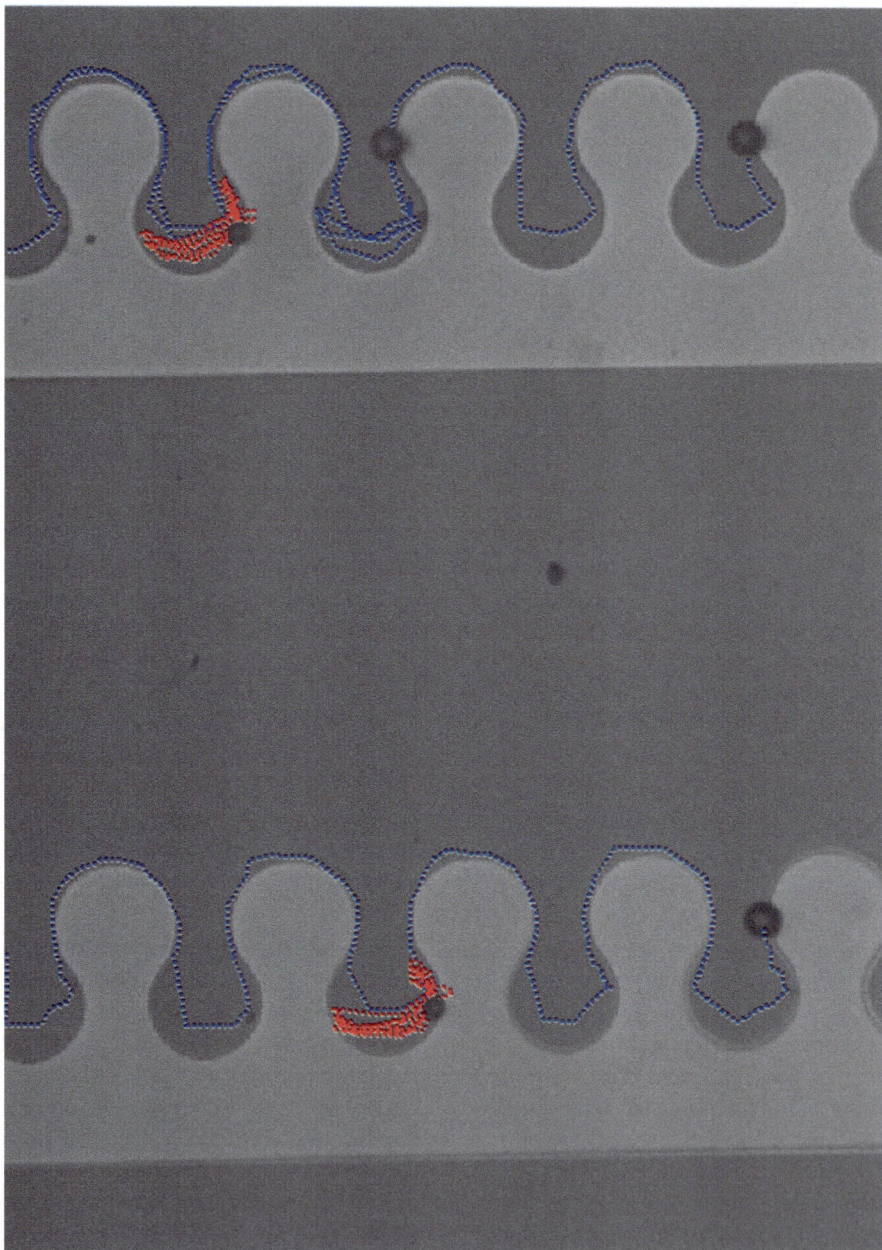

Fig. 7.31 Size-based magnetic particle separation using a drop-shaped magnetic conductor in a tri-axial magnetic field is shown. The magnetic field cone angle is $\alpha = 37°$, and the driving frequency is 0.1 Hz. The trajectories of the 5.6 µm and 8.4 µm diameter particles are depicted with red and blue dotted lines, respectively. The figure is taken with permission from [44]

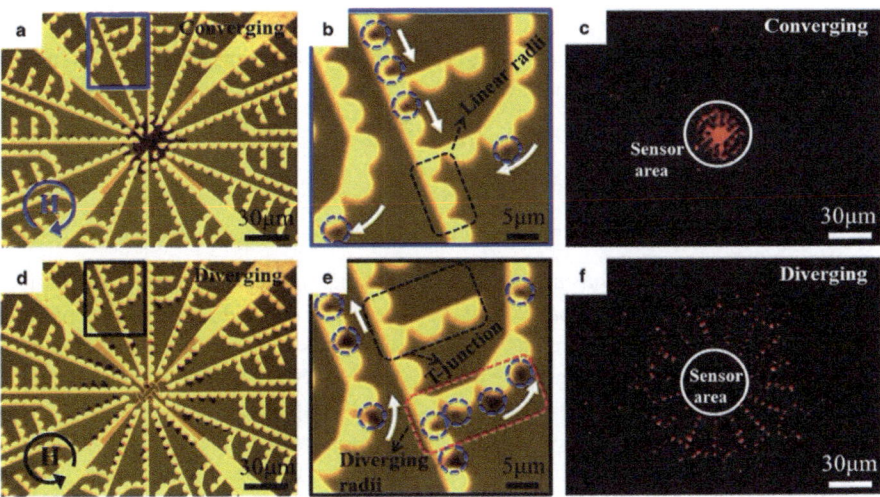

Fig. 7.32 Magnetophoretic circuit design for concentrating the magnetic particles is illustrated. (**a**) The particles move towards the center of the spider web at a clockwise rotating magnetic field. (**b**) A zoomed image of particle trajectories along the magnetic track is shown. The white arrow and the black dashed rectangle stand for the particle trajectories and the linear path around the junction, respectively. (**c**) A fluorescent image of the particles at the center is shown. The white circle depicts the sensor area. (**d**) The particles are dispersed away from the center at a counterclockwise rotating magnetic field. (**e**) A zoomed image of particle trajectories along the magnetic track is shown. The white arrow and the black and red dashed rectangles stand for the particle trajectories and the junctions which do not pass the particles (so they need to move along the curvy magnetic tracks), respectively. (**f**) A fluorescent image of the dispersed particles using the magnetophoretic circuit is shown. The figure is taken with permission from [45] under a Creative Commons Attribution 4.0 International License. http://creativecommons.org/licenses/by/4.0/

a good particle dispersion. The fluorescent image in Fig. 7.32**f** illustrates the particles along the dispersing tracks.

This design is successfully used for concentrating protein-coated superparamagnetic particles at the center of the circuit. The integrated PHR sensor at that spot detects the particles. The output voltage variation is proportional to the number of magnetic particles collected on the sensor. Hence, this circuit has the potentials in enhancing the sensor efficiencies with samples of low concentration.

To conclude, magnetophoretic circuit elements can be integrated to form useful circuits such as random access memories to form an array of single particles and cells. The memory architecture in these chips is similar to the ones in computer memories with the same addressing technique which lowers the number of required signaling wires. The magnetophoretic circuits are equipped with microfluidic chips to form magnetomicrofluidic circuits which are more reliable and can assemble particle arrays at much higher speeds, while it offers the same control and precision. In this method, the hydrodynamic particle trapping system is combined with the magnetophoretic circuits. After surface functionalization, monitoring and controlling the temperature and CO_2 concentration, and adjusting it at the right level, the chips are ready for running bio-experiments. In addition to single-cell analysis, magnetophoretic circuits are used in multiple other applications, such as biosensing, magnetometry, and particle separation.

7.6 Conclusions

Lab-on-a-chip technology has advanced various fields including cellular and molecular biology, at a tremendous pace. One main goal in this field is controlled particle transport. For example, to answer the complex questions buried within heterogeneous biological systems, the field is rapidly moving towards SCA platforms, which enable a reductionist approach to mechanistic biology. The proper SCA platform needs particle manipulation techniques to allow comprehensively phenotype and genotype single cells with sufficient automation, flexibility, and scale for studying the rare biological events relevant to the pathogenesis and outcome of human diseases.

Various methods based on magnetic, electric, acoustic, and hydrodynamic forces are developed. But the requirements mentioned above are answered by introducing the magnetomicrofluidic chips and magnetophoretic circuits. This novel method offers precise particle transport based on a new concept. In this technology, microfabrication methods, similar to the ones used in the electronics industry, are used. The operation of the device is also analogous to the one of the electronic circuits; however, the mobile components are single living cells and particles, as opposed to electrons. Hence, similar control and automation over a large number of particles in parallel are offered. In magnetomicrofluidic circuits, the magnetophoretic circuits are equipped with a microfluidic system, enabling them to transport the particles faster and enhancing the system efficiency.

The magnetomicrofluidic circuits can store a large number of individual cells as well as single-cell pairs into arrays and incubate them on the chip for multiple days. Two main advantages of this tool are (i) the ability to manipulate a large number of particles in parallel and (ii) having precise control over individual particles to place them at a particular spot on the chip. These important specifications make this tool unique with the potential ability to answer essential problems in biology and medicine.

Two types of magnetophoretic circuits based on the driving magnetic field exist. The simple form of these circuits operates in a 2D in-plane magnetic field. This type of circuit is based on serially connected magnetic disks or half-disks. But, in a pure in-plane magnetic field, the particles tend to agglomerate. The resulting particle clusters may alter the device's operation. Thus, magnetophoretic circuits operating in a tri-axial magnetic field, in which a vertical bias field is superimposed to the in-plane rotating field, are designed. The vertical field biases the particles such that the probability of particle cluster formation is dramatically lowered. Various magnetic patterns for these new circuits are offered. The drop-shape and the TI magnetic tracks are two good examples.

Providing a non-fouling layer on microchannel walls would prevent the cells from adhering to them and helps the magnetomicrofluidic chips for manipulating the single cells. This ability is demonstrated on both open chips (*i.e.*, magnetophoretic circuits) and the magnetomicrofluidic chips based on silicon and glass. The growth of polymers on these chips is well characterized. After polymerization, it is shown that the magnetic forces move the cells on the desired path.

To move the cells, they first need to be magnetized. The cells are labeled upon the principle of antibody-antigen binding between the nanoparticles and the cell. Also, methods based on cellular uptake (which can be enhanced by arginine-rich peptides) of magnetic nanoparticles are proposed.

Moreover, as an alternative magnetization technique, in situ labeling, in which the cells would be delivered in the hydrodynamic trap sites, and then magnetic nanoparticles would be introduced to label the cells, can be tried. Since in this method the cells are labeled after being trapped, it may reduce the possibility of nanoparticle loss.

The magnetomicrofluidic devices described in this text have the potentials to be used in important applications. For example, researchers can organize cancer or latently infected HIV single cells on the chip, apply different drugs to them, and study their behavior. They can also place sensing magnetic beads next to each cell to study their cytokine secretion profile in various situations over time. It is possible to put the barcode-carrying beads next to the cells, lyse them, and gather their transcriptome.

Moreover, it is possible to apply these capabilities to probe the interactions between single-cell pairs to advance applications in cell-based immunotherapies, cancer, and other diseases. For example, much remains unknown about the distance-dependent interactions between infected or cancer cell pairs, their signaling networks, and the cell phenotypes required for the efficient elimination of pathogenic invaders. By adapting the magnetomicrofluidic tool to study these and other biological questions, important insights about intercellular signaling networks and the communication hubs of biological operating systems can be revealed in future. This tool also has promise in developing targeted cancer therapies and in analyzing pathogen/host interactions where many open questions remain to be studied regarding the mechanisms for immunological protection against microbial pathogens.

In addition to SCA tools, these chips are utilized for detecting biomolecules (e.g., DNAs and proteins), magnetometry, particle separation, and so on. Also, the tool can be used for manipulating bioparticles, such as magnetotactic bacteria, with intrinsic magnetic properties and without the need for magnetic labeling. Assembling two sets of particles in magnetophoretic capacitors is illustrated. The first particle set can be cells of interest. The second set of particles can be a second cell type to study their interaction with the first type, magnetic barcode-carrying beads to carry genetic materials, magnetic bead sensors to detect cell cytokine secretions, or drug-releasing magnetic beads to perform drug studies on cells. These circuits provide us the ability to precisely define the location of these particles, remove any of them, or replace them with another particle as needed and in a programmable automatic fashion.

References

1. Slingerland, J. (2016). *Nanotechnology*. ABDO Publishing Company.
2. Hughes, J. M. (2015). *Practical Electronics: Components and Techniques*. O'Reilly Media.
3. Jackson, T. N. (2005). Organic sSemiconductors: Beyond Moore's Law. *Nature Materials, 4*(8), 581–582.
4. Abedini-Nassab, R., et al. (2015). Characterizing the sSwitching thresholds of magnetophoretic transistors. *Advanced Materials, 27*(40), 6176–6180.
5. Lim, B., et al. (2014). Magnetophoretic circuits for digital control of single particles and cells. *Nature Communications, 5*, 3846.
6. Abedini-Nassab, R. (2017). *Magnetomicrofluidics circuits for organizing bioparticle arrays.*
7. Halldorsson, S., et al. (2015). Advantages and challenges of microfluidic cell culture in polydimethylsiloxane devices. *Biosensors and Bioelectronics, 63*, 218–231.
8. Frimat, J. P., et al. (2011). A microfluidic array with cellular valving for single cell co-culture. *Lab on a Chip, 11*(2), 231–237.
9. Abedini-Nassab, R. (2019). Magnetomicrofluidic platforms for organizing arrays of single-particles and particle-pairs. *Journal of Microelectromechanical Systems, 28*(4), 732–738.
10. Abedini-Nassab, R., & Mahdaviyan, N. (2020). A microfluidic platform equipped with magnetic nano films for organizing bio-particle arrays and long-term studies. *IEEE Sensors Journal, 20*(17), 9668–9676.
11. Chary, S. R., & Jain, R. K. (1989). Direct measurement of interstitial convection and diffusion of albumin in normal and neoplastic tissues by fluorescence photobleaching. *Proc Natl Acad Sci U S A, 86*(14), 5385–5389.
12. Dafni, H., et al. (2002). Overexpression of vascular endothelial growth factor 165 drives peritumor interstitial convection and induces lymphatic drain: Magnetic resonance imaging, confocal microscopy, and histological tracking of triple-labeled albumin. *Cancer Research, 62*(22), 6731–6739.
13. Kingsmore, K. M., et al. (2018). MRI analysis to map interstitial flow in the brain tumor microenvironment. *APL Bioeng, 2*(3).
14. Wiig, H., & Swartz, M. A. (2012). Interstitial fluid and lymph formation and transport: Physiological regulation and roles in inflammation and cancer. *Physiological Reviews, 92*(3), 1005–1060.
15. Shemesh, J., et al. (2015). Flow-induced stress on adherent cells in microfluidic devices. *Lab on a Chip, 15*(21), 4114–4127.
16. Ingber, D. E. (1997). Tensegrity: The architectural basis of cellular mechanotransduction. *Annual Review of Physiology, 59*, 575–599.
17. Han, B., et al. (2004). Conversion of mechanical force into biochemical signaling. *Journal of Biological Chemistry, 279*(52), 54793–54801.
18. Abedini-Nassab, R. (2020). Magnetophoretic circuit biocompatibility. *Journal of Mechanics in Medicine and Biology, 20*(07), 2050050.
19. Yu, W., et al. (2014). A microfluidic-based multi-shear device for investigating the effects of low fluid-induced stresses on osteoblasts. *PLoS One, 9*(2), e89966.
20. Lu, H., et al. (2004). Microfluidic shear devices for quantitative analysis of cell adhesion. *Analytical Chemistry, 76*(18), 5257–5264.
21. Miltenyi, S., et al. (1990). High gradient magnetic cell separation with MACS. *Cytometry, 11*(2), 231–238.
22. Gao, H., et al. (2013). Ligand modified nanoparticles increases cell uptake, alters endocytosis and elevates glioma distribution and internalization. *Science and Reports, 3*, 2534.
23. Calero, M., et al. (2015). Characterization of interaction of magnetic nanoparticles with breast cancer cells. *J Nanobiotechnology, 13*, 16.
24. Reddy, L. H., et al. (2012). Magnetic nanoparticles: Design and characterization, toxicity and biocompatibility, pharmaceutical and biomedical applications. *Chemical Reviews, 112*(11), 5818–5878.

25. Wydra, R. J., et al. (2015). The role of ROS generation from magnetic nanoparticles in an alternating magnetic field on cytotoxicity. *Acta Biomaterialia, 25*, 284–290.
26. Nel, A., et al. (2006). Toxic potential of materials at the nanolevel. *Science, 311*(5761), 622–627.
27. Neamtu, M., et al. (2018). Functionalized magnetic nanoparticles: Synthesis, characterization, catalytic application and assessment of toxicity. *Science and Reports, 8*(1), 6278.
28. Schladt, T. D., et al. (2011). Synthesis and bio-functionalization of magnetic nanoparticles for medical diagnosis and treatment. *Dalton Transactions, 40*(24), 6315–6343.
29. Zebisch, K., et al. (2012). Protocol for effective differentiation of 3T3-L1 cells to adipocytes. *Analytical Biochemistry, 425*(1), 88–90.
30. Greig, R. G., et al. (1985). Tumorigenic and metastatic properties of "normal" and ras-transfected NIH/3T3 cells. *Proc Natl Acad Sci U S A, 82*(11), 3698–3701.
31. Vorrink, S. U., et al. (2014). Hypoxia perturbs aryl hydrocarbon receptor signaling and CYP1A1 expression induced by PCB 126 in human skin and liver-derived cell lines. *Toxicology and Applied Pharmacology, 274*(3), 408–416.
32. Vorrink, S. U., Hudachek, D. R., & Domann, F. E. (2014). Epigenetic determinants of CYP1A1 induction by the aryl hydrocarbon receptor agonist 3,3′,4,4′,5-pentachlorobiphenyl (PCB 126). *International Journal of Molecular Sciences, 15*(8), 13916–13931.
33. Abedini-Nassab, R., & Shourabi, R. (2022). High-throughput precise particle transport at single-particle resolution in a three-dimensional magnetic field for highly sensitive bio-detection. *Science and Reports, 12*(1), 6380.
34. Cantor, R., et al. (1993). A high performance integrated DC SQUID magnetometer. *IEEE Transactions on Applied Superconductivity, 3*(1), 1800–1803.
35. Grob, D. T., et al. (2018). Magnetic susceptibility characterisation of superparamagnetic microspheres. *Journal of Magnetism and Magnetic Materials, 452*, 134–140.
36. Min, C., et al. (2017). Integrated microHall magnetometer to measure the magnetic properties of nanoparticles. *Lab on a Chip, 17*(23), 4000–4007.
37. Sinha, B., et al. (2012). Micro-magnetometry for susceptibility measurement of superparamagnetic single bead. *Sensors and Actuators A: Physical, 182*, 34–40.
38. Liao, S. H., Huang, H. S., & Su, Y. K. (2019). Scaling characteristics of the magnetization increments of functionalized nanoparticles determined using a vibrating sample magnetometer for liquid magnetic immunoassays. *IEEE Sensors Journal, 19*(20), 8990–8994.
39. Balk, A. L., et al. (2015). Quantitative magnetometry of ferromagnetic nanorods by microfluidic analytical magnetophoresis. *Journal of Applied Physics, 118*(9), 093904.
40. Abedini-Nassab, R., Ding, X., & Xie, H. (2022). A novel magnetophoretic-based device for magnetometry and separation of single magnetic particles and magnetized cells. *Lab on a Chip, 22*(4), 738–746.
41. Fuh, C. B., et al. (2000). A method for determination of particle magnetic susceptibility with analytical magnetapheresis. *Analytical Chemistry, 72*(15), 3590–3595.
42. Zborowski, M., et al. (1995). Analytical magnetapheresis of ferritin-labeled lymphocytes. *Analytical Chemistry, 67*(20), 3702–3712.
43. Yoon, J., et al. (2022). Magnetophoretic micro-distributor for controlled clustering of cells. *Adv Sci (Weinh), 9*(6), e2103579.
44. Abedini-Nassab, R., et al. (2016). Magnetophoretic conductors and diodes in a 3D magnetic field. *Advanced Functional Materials, 26*(22), 4026–4034.
45. Lim, B., et al. (2017). Concentric manipulation and monitoring of protein-loaded superparamagnetic cargo using magnetophoretic spider web. *NPG Asia Materials, 9*(3), e369–e369.

Printed by Printforce, the Netherlands